Green Chemistry: An Introduction

Green Chemistry: An Introduction

Green Chemistry: An Introduction

Barney Cooper

MURPHY & MOORE
www.murphy-moorepublishing.com

Published by Murphy & Moore Publishing
1 Rockefeller Plaza,
New York City, NY 10020, USA
www.murphy-moorepublishing.com

Green Chemistry: An Introduction
Barney Cooper

International Standard Book Number: 978-1-63987-270-1 (Hardback)

Cataloging-in-Publication Data

Green chemistry : an introduction / Barney Cooper.
p. cm.
Includes bibliographical references and index.
ISBN 978-1-63987-270-1
1. Green chemistry. 2. Environmental chemistry. 3. Chemical engineering.
4. Chemistry, Technical. 5. Sustainable engineering. I. Cooper, Barney.
TP155.2.E58 G74 2022
660--dc23

TABLE OF CONTENTS

PREFACE

It is with great pleasure that I present this book. It has been carefully written after numerous discussions with my peers and other practitioners of the field. I would like to take this opportunity to thank my family and friends who have been extremely supporting at every step in my life.

Green chemistry is concerned with the study of designing of processes and products that reduce or eliminate the formation and use of hazardous substances. It deals with the environmental impact of chemistry. It is also referred to as sustainable chemistry as it deals with the problems of resource depletion and chemical pollution. The goal of this discipline is to be more resource efficient by finding ways to reduce consumption of these resources and formulate technological approaches to prevent pollution. Atom economy, use of renewable feedbacks, real-time analysis for pollution prevention, less hazardous chemical synthesis, inherently safer chemistry for accident prevention, design for degradation and design for energy efficiency are some of the principles on which the discipline operates. This book elucidates the concepts and innovative models around prospective developments with respect to this discipline. It is compiled in such a manner, that it will provide an in-depth knowledge about the theory and practice of green chemistry. Through this book, we attempt to further enlighten the readers about the new concepts in this field.

The chapters below are organized to facilitate a comprehensive understanding of the subject:

Chapter - What is Green Chemistry?

The branch of chemistry that deals with the designing of processes and products to mitigate the use of hazardous substances is known as green chemistry. It applies across the life cycle of a chemical product, including its design, manufacture, use, and ultimate disposal. This is an introductory chapter which will briefly introduce about green chemistry.

Chapter - Principles and Concepts of Green Chemistry

Green chemistry matrics and control keys, waste management and minimization, atom economy, addition reaction, witting reaction, elimination reaction, substitution reaction, etc. are some of the principles and concepts studied under green chemistry. The topics elaborated in this chapter will help in gaining a better perspective of the principles and concepts of green chemistry.

Chapter - Green Synthesis and Chemistry

Green synthesis provides environmental and economic benefits through the use of non-toxic and safe reagents. It avoids the production of unwanted or harmful by-products through the build-up of reliable, sustainable, and eco-friendly synthesis procedures. This chapter discusses the subject of green synthesis and chemistry in detail.

Chapter – Catalysis in Green Chemistry

Catalysis is one of the fundamental principles of green chemistry that produces substantial material and saves energy. A few of its aspects are biocatalysis, enzyme catalysis, heterogenous catalysis, homogenous catalysis, photocatalysis, etc. This chapter has been carefully written to provide an easy understanding of catalysis in green chemistry.

Chapter – Green Reagent and Green Solvent

Green reagent deals with the reduction or elimination of the use or generation of hazardous substances in the design, manufacture and application of chemical products. The biosolvents that are derived from the processed agricultural crops are called green solvent. All the aspects related to green reagent and solvent have been carefully analyzed in this chapter.

Chapter – Green Chemistry and the Environment

Green chemistry aids in reducing pollution by minimizing the hazards of chemical feedstocks, solvents and products. Remediation removes hazardous materials from the environment and maintains environmental sustainability. This chapter closely examines the role of green chemistry in the environment for an extensive understanding of the subject.

Chapter – Applications

Green chemistry has a wide range of applications in agriculture, medical science and industries. Its major applications include the preservation and maintenance of environment sustainability, using agrochemicals, etc. This chapter sheds light on the various applications of green chemistry for a thorough understanding of the subject.

Barney Cooper

What is Green Chemistry?

1

- **Green Chemistry in Day-to-day Life**
- **Potential of Green Chemistry**
- **Green Toxicology**

The branch of chemistry that deals with the designing of processes and products to mitigate the use of hazardous substances is known as green chemistry. It applies across the life cycle of a chemical product, including its design, manufacture, use, and ultimate disposal. This is an introductory chapter which will briefly introduce about green chemistry.

Green chemistry is the design of chemical products and processes that reduce or eliminate the use or generation of hazardous substances. Green chemistry applies across the life cycle of a chemical product, including its design, manufacture, use, and ultimate disposal. Green chemistry is also known as sustainable chemistry.

- Prevents pollution at the molecular level.
- Is a philosophy that applies to all areas of chemistry, not a single discipline of chemistry.
- Applies innovative scientific solutions to real-world environmental problems.
- Results in source reduction because it prevents the generation of pollution.
- Reduces the negative impacts of chemical products and processes on human health and the environment.
- Lessens and sometimes eliminates hazard from existing products and processes.
- Designs chemical products and processes to reduce their intrinsic hazards.

How Green Chemistry Differs from Cleaning up Pollution?

Green chemistry reduces pollution at its source by minimizing or eliminating the hazards of chemical feedstocks, reagents, solvents, and products. This is unlike cleaning up pollution (also called remediation), which involves treating waste streams (end-of-the-pipe treatment) or cleanup of environmental spills and other releases. Remediation may include separating hazardous chemicals from other materials, then treating them so they are no longer hazardous or concentrating them for safe disposal. Most remediation activities do not involve green chemistry. Remediation removes hazardous materials from the environment; on the other hand, green chemistry keeps the hazardous materials out of the environment in the first place.

If a technology reduces or eliminates the hazardous chemicals used to clean up environmental contaminants, this technology would qualify as a green chemistry technology. One example is replacing a hazardous sorbent chemical used to capture mercury from the air for safe disposal with an effective, but nonhazardous sorbent. Using the nonhazardous sorbent means that the hazardous sorbent is never manufactured and so the remediation technology meets the definition of green chemistry.

Green Chemical Reactions

Reactions play major role in synthesis. The thought of Green Chemistry appeals for the development of new chemical reactivity's and reactions that can potentially provide benefits for chemical syntheses in understandings of resource and energy efficiency, product selectivity, operational simplicity, and health and environmental safety. Some of green reaction methods include atom economy where the reaction seeks to maximize the incorporation of the starting materials into the final product of any given reaction. In bio-catalysis of usefulness in various catalysts such as enzymes, whole cells, and antibodies for organic synthesis which have become more recognized.

- Aqueous Phase reactions.
- Biocatalysts in Organic Synthesis.
- Safer Reagents for Synthesis.
- Microwave Induced Green Synthesis.
- Organic Synthesis in Solid State.
- Green Reaction Media and Related Green Initiatives.
- Phase-Transfer Catalysis in Green Synthesis.
- Ultrasound Assisted Green Synthesis.

Benefits of Green Chemistry

Human Health

- Cleaner air: Less release of hazardous chemicals to air leading to less damage to lungs.

- Cleaner water: Less release of hazardous chemical wastes to water leading to cleaner drinking and recreational water.
- Increased safety for workers in the chemical industry: Less use of toxic materials, less personal protective equipment required, less potential for accidents (e.g., fires or explosions).
- Safer consumer products of all types: New, safer products will become available for purchase; some products (e.g., drugs) will be made with less waste; some products (i.e., pesticides, cleaning products) will be replacements for less safe products.
- Safer food: Elimination of persistent toxic chemicals that can enter the food chain; safer pesticides that are toxic only to specific pests and degrade rapidly after use.
- Less exposure to such toxic chemicals as endocrine disruptors.

Environment

- Many chemicals end up in the environment by intentional release during use (e.g., pesticides), by unintended releases (including emissions during manufacturing), or by disposal. Green chemicals either degrade to innocuous products or are recovered for further use.
- Plants and animals suffer less harm from toxic chemicals in the environment.
- Lower potential for global warming, ozone depletion, and smog formation.
- Less chemical disruption of ecosystems.
- Less use of landfills, especially hazardous waste landfills.

Economy and Business

- Higher yields for chemical reactions, consuming smaller amounts of feedstock to obtain the same amount of product.
- Fewer synthetic steps, often allowing faster manufacturing of products, increasing plant capacity, and saving energy and water.
- Reduced waste, eliminating costly remediation, hazardous waste disposal, and end-of-the-pipe treatments.
- Allow replacement of a purchased feedstock by a waste product.
- Better performance so that less product is needed to achieve the same function.
- Reduced use of petroleum products, slowing their depletion and avoiding their hazards and price fluctuations.

- Reduced manufacturing plant size or footprint through increased throughput.
- Increased consumer sales by earning and displaying a safer-product label (e.g., Safer Choice labeling).
- Improved competitiveness of chemical manufacturers and their customers.

Green Chemistry in Day-to-day Life

Green Dry Cleaning of Clothes

Perchloroethylene (PERC), $Cl_2C=CCl_2$ is commonly being used as a solvent for dry cleaning. It is now known that PERC contaminates ground water and is a suspected carcinogen. A technology, known as Micell technology developed by Joseph De Simons, Timothy Romark, and James McClain made use of liquid CO_2 and a surfactant for dry cleaning clothes, thereby replacing PERC. Dry cleaning machines have now been developed using this technique. Micell Technology has also evolved a metal cleaning system that uses CO_2 and a surfactant thereby eliminating the need of halogenated solvents.

Versatile Bleaching Agents

It is common knowledge that paper is manufactured from wood (which contains about 70% polysaccharides and about 30% lignin). For good quality paper, the lignin must be completely removed. Initially, lignin is removed by placing small chipped pieces wood into a bath of sodium hydroxide (NaOH) and sodium sulphide (Na_2S). By this process about 80-90% of lignin is decomposed. The remaining lignin was so far removed through reaction with chlorine gas (Cl_2). The use of chlorine removes all the lignin (to give good quality white paper) but causes environmental problems. Chlorine also reacts with aromatic rings of the lignin to produce dioxins, such as 2,3,4-tetrachlorop-dioxin and chlorinated furans. These compounds are potential carcinogens and cause other health problems.

These halogenated products find their way into the food chain and finally into products, pork, beef and fish. In view of this, use of chlorine has been discouraged. Subsequently, chlorine dioxide was used. Other bleaching agents like hydrogen per oxide (H_2O_2), ozone (O_3) or oxygen (O_2) also did not give this the desired results. A versatile agent has been developed by Terrence Collins of Camegie Mellon University. It involves the use of H_2O_2 as a bleaching agent in the presence of some activators known as TAML activators10 that as catalysts which promote the conversion of H_2O_2 into hydroxyl radicals that are involved in oxidation (bleaching). The catalytic of TAML activators allow H_2O_2 to break down more lignin in a shorter time and at much lower temperature. These bleaching agents find use in laundry and results in lesser use of water.

Green Solution to Turn Turbid Water Clear

Tamarind seed kernel powder, discarded as agriculture waste, is an effective agent to make municipal and industrial waste water clear. The present practice is to use Al-salt to treat such water. It has been found that alum increases toxic ions in treated water and could cause diseases like Alzheimer's. On the other hand kernel powder is not-toxic and is biodegradable and cost effective. For the study, four flocculants namely tamarind seed kernel powder, mix of the powder and starch, starch ad alum were employed. Flocculants with slurries were prepared by mixing measured amount of clay and water.

The result showed aggregation of the powder and suspended particles were more porous and allowed water to ooze out and become compact more easily and formed larger volume of clear water. Starch flocks on the other hand were found to be light weight and less porous and therefore didn't allow water to pass through it easily. The study establishes the powder's potential as an economic flocculants with performance close more established flocculants such as $K_2SO_4Al_2(SO_4)_3.24H_2O$ (potash alum).

Potential of Green Chemistry

Oxidation Chemical Agent and Contact Action

Several of the oxidization reagents and catalysts are comprised of nephrotoxic substances like significant metals. Since these substances were typically employed in very massive volumes needed to convert numerous pounds of petrochemicals, there was a major inheritance of those metals being discharged to the setting and having substantial negative impact on human health and setting. It may be modified by the utilization of benign substances.

Biometric Multifunctional Reagents

Whereas artificial contact action and reagents for the foremost half have targeted on concluding one distinct transformation. The manipulations could embrace activation, conformational changes, and one or many actual transformations and derivitizations.

Combinatorial Inexperienced Chemistry

The chemistry of having the ability to create massive numbers of chemical compounds chop-chop on a little scale exploitation reaction matrices. The instance is lead that incorporates a massive no of derivatives. This chemistry has enabled massive no of gear to be created and their properties assessed while not the magnitude of the consequences of waste disposal.

Energy Focus

The environmental impact of energy usage square measure profound however haven't been as visible and as direct as a number of the hazards that haven't been expose by materials employed in manufacture, use and disposal of chemicals. The advantage of contact action is dramatic in chemical science. there's a requirement to style substances and materials that square measure effective, economical and cheap at the capture, storage and transportation.

Proliferation of Solvent Less Reactions

One in every of the 'solvent alternatives' that's being: it's one in every of the solvent alternatives that's being developed in inexperienced chemistry is that of solvent less reaction system. The carrying of producing method in solvent-less condition utilizes some non-traditional conditions. This helps in development of product isolation, separation and purification that may be solventless yet so as to maximise.

Supramolecular Chemistry

Analysis is presently current within the space of supramolecular chemistry to generate reactions which may proceed in the solid phase without use of solvents. The cyclic addition of trans-1,2- bis(4-pyridyl)ethylene is directed by phenol within the solid state.

Green Toxicology

Green Toxicology refers to the application of predictive toxicology in the sustainable development and production of new less harmful materials and chemicals, subsequently reducing waste and exposure. Built upon the foundation of "Green Chemistry" and "Green Engineering", "Green Toxicology" aims to shape future manufacturing processes and safe synthesis of chemicals in terms of environmental and human health impacts. Being an integral part of Green Chemistry, the principles of Green Toxicology amplify the role of health-related aspects for the benefit of consumers and the environment, in addition to being economical for manufacturing companies. Due to the costly development and preparation of new materials and chemicals for market entry, it is no longer practical to ignore the safety and environmental status of new products during product development stages. However, this is only possible if toxicologists and chemists work together early on in the development of materials and chemicals to utilize safe design strategies and innovative in vitro and in silico tools.

The use of solvents, which consistently account for approximately 80–90% of the materials used in a typical pharmaceutical batch chemical operation, may play a dominant

Integration of Green Toxicology in Discovery, Development and Production Practices

In order to efficiently develop new compounds or products with the desired technological or biological traits and lesser toxicity, many different structures need to be evaluated. Synthesizing these new molecules, often with complex chemical structures, in sufficient amounts is very demanding. The discovery and development of new products and active ingredients (AIs) thus rely on reaction screening and route scouting with high-throughput experimentation incorporating automated solid and liquid handling for rapid and routine screening (e.g. evaluate typically 100 reactions simultaneously at 1 mL scale). Principle design is incorporated in all aspects of the AI process, including reaction screening, optimization, critical parameter identification and process response surface modelling; however, the toxicity of all steps of chemical development and processing (e.g. intermediates, solvents, catalysts, etc.) must be considered.

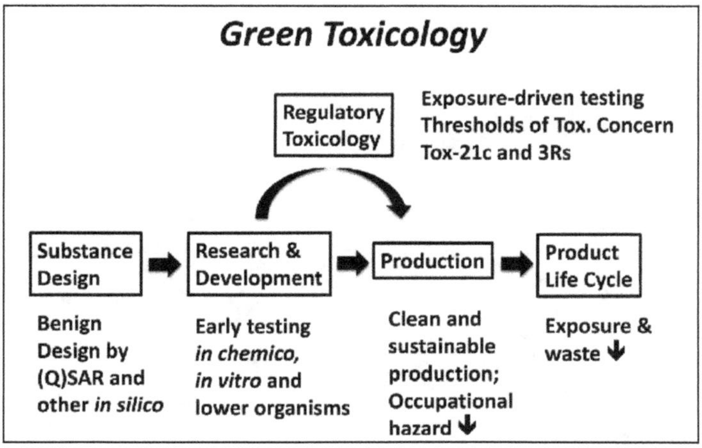

Identifying the anticipated biological traits of AIs and products often requires only one assay, as the target is known. In contrast, the "off-target" specification (reduced toxicity) requires many tests, including those with animals. The reduction of animal testing is another pillar of ethical-based sustainability. Animal testing enables the assessment of toxic effects and thus helps to develop less toxic compounds. Animal testing is actually performed rather late during the development of new chemicals. There are a number of other negative aspects associated with animal testing including animal welfare and ethics, as well as the heavy reliance and use of time (months to years per assay), money (thousands to millions of dollars per testing programme) and resources. In addition, the uncertainty associated with the extrapolation from animals to humans may be accounted for using uncertainty or safety factors; however, many variables such as body weight, homology of genetic material and enzymes, as well as anatomy and physiology of organisms and humans must be considered to make confident conclusions about human hazard assessments from animal data.

Animal testing will, for the foreseeable future, be an essential and compulsory step in assessing the risk and regulation of chemical development, as various industrial sectors have different legal constraints placed upon them as consequences of registration and approval procedures worldwide. However, new tools and Green Toxicology strategies to relate mechanistic information obtained through in silico, toxicogenomic and in vitro approaches for animal and human effects have been investigated as alternative or complementary tools for in vivo tests. For example, with the development of alternative in vitro methods, test substance demand is generally reduced to less than 500 mg per assay. Hence, with only a few grams of a new compound, many tests can be performed.

Principles and Concepts of Green Chemistry

2

- Green Chemistry Control Keys
- Green Chemistry Matrics
- Waste Management
- Waste Minimisation
- Atom Economy
- Addition Reaction
- Substitution Reaction
- Elimination Reaction

Green chemistry matrics and control keys, waste management and minimization, atom economy, addition reaction, witting reaction, elimination reaction, substitution reaction, etc. are some of the principles and concepts studied under green chemistry. The topics elaborated in this chapter will help in gaining a better perspective of the principles and concepts of green chemistry.

Green chemistry is generally based on the 12 principles proposed by Anastas and Warner. Nowadays, these twelve principles of green chemistry are considered the fundaments to contribute to sustainable development. The principles comprise instructions to implement new chemical products, new synthesis, and new processes.

The twelve principles of green chemistry proposed by anastas and warner:

- The "better to prevent than to cure" principle: It is beneficial to a priori prevent the generation of waste instead of later on treating and cleaning up waste.

- The "atom economy" principle: Synthetic production routes have to be planned in a way maximizing the incorporation of all the compounds used in the synthesis into the desired product.
- The "less precarious chemical syntheses" principle: Wherever feasible, such synthetic methods have to be aspired, which resort to and generate compounds of no or only insignificant noxiousness to the environment and human health.
- The "designing safer chemicals" principle: Chemicals should be developed in a way affecting their desired functionality, while, at the same time, considerably reducing their toxicity.
- The "safer solvents and safer auxiliaries" principle: Expenditure of auxiliary substances, such as solvents, separation agents, and others, should be avoided wherever possible; if not possible, harmless auxiliaries should be used.
- The "design for energy efficiency" principle: The environmental and economic impact of energy demands for chemical processes should be analyzed in terms of followed by optimizing the required energy input. Wherever practicable, chemical synthesis should be carried out under mild process conditions, hence, at ambient temperature and pressure.
- The "design for energy efficiency" principle: The environmental and economic impact of energy demands for chemical processes should be analyzed in terms of followed by optimizing the required energy input. Wherever practicable, chemical synthesis should be carried out under mild process conditions, hence, at ambient temperature and pressure.
- The "renewable feedstocks" principle: Whenever feasible in technological and economic terms, synthetic processes should resort to such raw materials and feedstocks, which are renewable rather than limited.
- The "renewable feedstocks" principle: Whenever feasible in technological and economic terms, synthetic processes should resort to such raw materials and feedstocks, which are renewable rather than limited.
- The "derivative reduction" principle: Redundant derivatization, e.g., protection/deprotection, the use of blocking groups, or temporary modification of physical/chemical processes, requires additional reagents and often contributes to additional waste generation. Therefore, wherever possible, they should be avoided or reduced to a minimum.
- The "catalysis" principle: Generally, catalytic reagents are intrinsically superior to stoichiometric reagents; these catalysts should be as selective as possible.
- The "degradation" principle: Chemical products have to be designed in such a way that, at the end of their life span, they do not resist in the biosphere, but disintegrate into nontoxic degradation products.

- The "real-time analysis for pollution prevention" principle: Advanced analytical methods have to be developed, which permit the real-time, in-line process monitoring and control well before hazardous substances are generated.
- The "accident prevention by inherently safer chemistry" principle: Compounds and the compound's formula applied in a chemical process should be chosen in a way minimizing the risk of chemical accidents, encompassing the release of chemicals, detonations, or fire formation.

"It is Better to Prevent Waste than to Treat or Clean up Waste after its Formation."

This statement is one of the most popular guidelines in process optimization; it describes the ability of chemists to redesign chemical transformations in order to minimize the generation of hazardous waste as an important step toward pollution prevention. By preventing waste generation, the hazards associated with waste storage, transportation, and treatment could be minimized.

This principle is easy to understand and easy to apply, and examples from both industry and academia have proven its significance, relevance, and feasibility. This pillar of green chemistry is still valid; however, we have to conceive it in a broader context, switching from a restricted interpretation of waste based on its quantity to a universal approach to deal with the topic "waste":

- We have to take waste's multidimensional nature into account.
- We need to move from a "quantity of waste per quantity of product" principle toward a principle addressing the "quantity of waste generated per function provided by the product." In this context, we have to aim at making both quality and functionality of products superior.
- Considering the entire life cycle of a product, we have to address the fact that not only the production process itself generates waste but, moreover, "end-of-life waste" accrues after the product's life span or its consumption. This encompasses firstly the conversion of such materials up to now considered as waste into valuable products and, secondly, their recyclability.

Generally, moving toward "zero-waste production" and "waste prevention" encompasses the modernization of industrial processes through clean production techniques. These techniques aim at the reduction of gaseous emissions, effluents, solid residues, and noise generation; generally, they are developed to contribute to the protection of climate and environment. However, the most auspicious strategy to prevent waste generation would simply be not producing the desired product. In most scenarios, this will not be practicable; however, it might be reasonable to instead produce completely novel products, which display higher quality and longer durability. Lower quantities of such novel, superior products are sufficient to fulfill a desired function. An alternative approach is to avoid that the product can be transformed into precarious waste, e.g., by making plastics accessible toward biodegradation or by a priori switching toward biodegradable plastic instead of highly recalcitrant petrochemical plastics. According to these ideas, we need to fundamentally reconsider our understanding of waste as hazardous material that needs to be disposed by enhancing the status of waste to a valued resource, which can act as starting material for generation of new products.

Maximize Atom Economy

Atom economy is a concept developed in the early 1990s to evaluate the efficacy of chemical conversions on an element-by-element basis. In analogy to well-established yield calculations, the concept of "atom economy" is based on the ratio of the entire mass of atoms in the target product to the entire mass of atoms in the starting materials. One option to reduce waste generation is to plan such chemical transformations, which maximize the integration of all materials used in the process into the final product, resulting in a number of wasted atoms as low as possible. Hence, selecting such chemical conversion routes, which incorporate the major share of starting materials into desired products, displays higher efficiency and contributes to waste reduction. This concept is nowadays widely implemented in new routes to generate various organic compounds, e.g., such substances that are used in the biomedical and pharmaceutical field.

One factor that is greatly speeding the incorporation of pollution prevention into industrial manufacturing processes is the development of green chemistry. According to an alternative definition, chemical synthesis methods should be designed in such a way to maximize the incorporation of different potentially hazardous materials, such as spinning precarious waste with cement and sand to produce improved paste used in construction applications or to incorporate radioactive wastes as immobilizing material to produce a safe stabilized form of waste. In a similar vein, combining recycled poly(ethylene terephthalate) (PET) waste with cement paste displays a viable strategy for immobilization of hazardous wastes, e.g., radioactive borates.

Design Less Hazardous Chemical Synthesis

In synthetic organic chemistry, effecting a successful chemical transformation in a new way or with a new molecule or in a new order is what matters regarding the principles of green chemistry. Various researchers have clearly demonstrated the direct relation of toxicity and the associated hazards and risks allied with chemical reactions to the matrix of matter present in the reaction vessel. Generally, the holistic toxicity spectrum of products or processes, together with most other sustainability and green chemistry criteria, is highly impacted by the chemistry behind a process and the transformation contributing to a chemical synthesis chain. An exception is identified in such cases where a molecule is produced by purpose, which is designed to display toxicity and/or biologically activity. For example, this scenario is found in the case of various molecules synthesized for pharmaceutical or agricultural applications; such compounds exhibit toxicity and/or impact living organisms.

Selection of compounds and materials to be used to increase the efficacy of chemical transformations is a pivotal point in process development; chemists should dedicate increased attention to the decision on which materials to be put into reaction vessels. It is simple to disregard all the other materials and to dedicate all efforts exclusively to the chemosynthetic pathway, which provides us with the desired product. However, discounting all the other matter present in a production process ultimately results in a high price to be payed, and we finally have to get rid of this scenario. Sometimes, chemists actually produce hazardous molecules, and, therefore, the subsequent principle is dedicated to the design of molecules which are intrinsically safer in their nature.

Design Safer Chemicals and Products

Chemical products should be designed to achieve their desired function with at the same time minimizing their toxicity. New products can be designed that are inherently safer, while highly effective for the target application. For example, the direct incorporation of radioactive spent liquid scintillation waste into cement combined with clay materials is considered an added value in the immobilization of the hazardous organic wastes in very cheap materials and natural clay to produce a safe stabilized product easy for handling, transformation, and disposal.

Safer Solvents and Auxiliaries

This principle promotes the use of safer solvents and auxiliaries. It is about any substances that do not directly contribute to the structure of the reaction product but are still necessary for the chemical reaction or process to occur. Mostly, reactions of organic compounds take place in liquid milieus, where the solvent acts in different ways: it can enable enhanced contact between the reactants, it can stabilize or destabilize generated intermediates, or it can influence transition states. In addition, the applied solvent also governs the selection of adequate downstream and regeneration processes and recycling or discarding techniques. By taking the ecological effect of chemical processes in consideration, innovative concepts for substitution of volatile organic solvents have become a great challenge in green chemistry. A green solvent should meet numerous criteria such as low toxicity, nonflammability, nonmutagenicity, nonvolatility, and widespread availability among others. Moreover, these green solvents have to be cheap and easy to handle and recycle. Prime examples are provided in the field of extractive recovery of microbial polyhydroxyalkanoate (PHA) biopolymers, typical intracellular storage materials, from biomass . For this extraction process, which is typically carried out by the use of precarious halogenated solvents, one more and more resorts to less harmful greener solvents, or to new recovery methods which entirely do without any solvents.

Design for Energy Efficiency

Usually, energy is used to enhance the human life in important ways. The traditionally used energy sources like coal, oil, and gas are limited in supply, and their combustion releases greenhouse gases. For continuous improvement of life quality, both move toward renewable energy and design for energy efficiency are needed. Designing more efficient processes by choosing the most suitable technologies and unit operations has to go in parallel with selecting proper energy sources. Using an electric motor with energy sources generated from the sun and wind is more effective in ecological terms instead of using fossil fuels. How energy is converted to useful forms and where it gets lost are the most important questions for engineers and designers to help society use energy more effectively.

Consequently, green chemistry includes minimization of energy loss like mechanical friction, fluid drag, and unwanted heat transfer, by improving the layout and insulation of a refrigerator or designing lighter vehicles with enhanced aerodynamic characteristics and lower rolling resistance.

In addition, when developing a new production process, the effect of geographical location of production plants has to be taken into account: ecological comparison of different production scenarios for the same product, in this case bioplastics, clearly demonstrates that different energy production technologies, resources for energy production, and the effect of available energy mixes in different countries become significant for the ecological footprint of a new process.

Use of Renewable Feedstocks

According to the principles of green chemistry, a raw material or feedstock should be renewable rather than depleting whenever technically and economically practicable. Using renewable resources like microbial or plant biomass, which are embedded into nature's closed carbon cycle, represents a real option to prepare functional bioproducts in a sustainable way and to contribute to energetic transition.

In the context of the Green Chemistry the renewable feedstock thematic, we nowadays witness a vast number of current multidisciplinary collaborations, involving the fields of, among others, agronomy, biochemical engineering, biotechnology, chemistry, microbiology, physics, toxicology, or engineering. These collaborations result in the development of next-generation fuels, polymers, and other materials pivotal for our today's society based on renewable resources and characterized by low impact on health and environment. The current global dynamic of these developments indeed gives reason to optimism for the future. Finding a method to convert raw wastes such as spinney waste fibers into a mortar composite stabilized material could be an excellent application of this principal of green chemistry. Whenever switching from fossil feedstocks to renewables, one has to consider that using renewable resources enlarges the process concept by incorporating resource provision, transportation, storage, and other aspects of logistics into the process design. Such a switch in feedstock provision, however, results in a fundamental change in the structure of processes, used technologies, and the economical concepts of industry and society.

Reduce Derivatives

Many processes could be designed in such a way to reduce the use of additional reagents and the resulting wastes. It is commonly necessary to synthesize a derivative of a compound containing groups which are not needed in the final product, but which allow the synthesis or purification steps to proceed more easily. However, these derivatives result in lower atom economy, since they introduce atoms that are not incorporated into the final product, but finally end up as waste; this is in conflict with the atom efficiency principle. For many reactions that have traditionally required protecting groups, chemists are currently devoting research efforts to finding alternatives that do without them.

Catalysis

Catalysis is the chemical reaction enabled or accelerated by a catalyst. According to Ostwald, catalysts are substances that speed up a reaction by enabling an energetically favored transition state between reactants, but which are not consumed by it and do not appear in the net reaction equation. Catalysts play an essential role in our modern industrial economy, in our stewardship of the environment, and in all biological processes. Saleh and others found that iron and copper sulfate as catalysts improved the mechanism of oxidative degradation of cellulosic wastes using 35% hydrogen peroxide. High weight reduction up to 95.2% in the presence of copper sulfate and 87.8% in the presence of ferrous sulfate was reported. Recently, synthesis of nano-catalysts of specific size and shape was developed to allow facile movement of materials in the reacting phase and the control over morphology of nanostructures to tailor their physical and chemical properties. Nano-catalyst systems encompassing a paramagnetic core allow rapid and selective chemical transformations with excellent product yield coupled with the ease of catalyst separation and recovery.

Talking about catalysis, it is nowadays indispensable to spend some words on the topic biocatalysis; hence, the application of enzymes as highly selective and active catalysts produced by Mother Nature. Not only do enzymes carry out the desired reactions under mild conditions of temperature and pressure, which is analogous to the energy efficiency principle. Moreover, they are predestined to drive reactions of renewable materials (analogy to the renewable feedstock principle) and, in some cases, even enable reactions that are not accessible by using traditional catalysts, such as the generation of some enantio-pure products. In addition, biocatalysts in a free or immobilized form are to an increasing extent applied for bioremediation, hence, the mitigation of pollutants from the ecosphere.

In the context of catalysis, microwave-driven chemical reaction for organic chemistry is an emerging field to drastically reduce reaction times from days to only seconds. During the last decades, this technique has matured from a laboratory shenanigan to a well-established technology, now already industrially applied.

Design for Degradation

One of the most important objectives of green chemistry is maximizing the production with minimizing unwanted by-products. Designing of products and processes that display reduced impact on humans and the environment, such as creating sustainable mortar composite that could be considered as an value-added product suitable for various applications as inert matrix for immobilization of some low and intermediate levels of radioactive wastes, decorative tiles, building bricks, and light concrete, is reported. In this case, highly reactive hydroxyl radicals react with the organic moieties of the spinning fiber wastes either by subtracting ions of hydrogen or by addition to the unsaturated site to yield organic radicals, which are readily oxidized by oxygen. Therefore, the end products of the degradation process were only carbon dioxide and water.

Real-time Analysis for Pollution Prevention

With advancements in chemistry, the production of numerous toxic chemicals is a serious problem for the environment. One of the basic concepts of green chemistry is familiar to pollution prevention practitioners. Less hazardous materials in chemical formulations and reducing waste formation have been sounded for many years. Consequently, green chemistry aims at eliminating the usage and generation of hazardous substances by designing better manufacturing processes for chemical materials with minimum waste production by real-time monitoring of running processes. This consequently enables a timely intervention right before waste or toxins are generated.

Inherently Safer Chemistry for Accident Prevention

It is of outstanding importance to avoid highly reactive chemicals that could potentially cause accidents during the reaction. Substance and the form of a substance used in a chemical process should be chosen in such a way to minimize the potential for chemical accidents, including toxin releases, explosions, and fire formation. For example, the most abundant solution medium, water, could accidentally cause an explosion by flowing into a tank containing methyl isocyanate gas, releasing a large amount of methyl isocyanate into the surrounding area. Other well-known materials, which undergo reactions of often disastrous outcome with water, are found among alkali metals. If an alternative reaction had been developed that did not use this reagent, the risk of explosion even causing death would have been minimized.

Intrinsically, safe chemistry can also be carried out in flow mode, using tubular microreactors with reaction channels of tiny diameter. Such flow chemistry approaches drastically reduce the reaction volume, the reaction time, and catalyst requirement, intensifies the processes by boosting the space/time yield, opens new process windows in terms of extreme temperature and pressure conditions to be applied, and, moreover, even allows to carry out highly dangerous reactions in a safe way. In addition, the application of flow chemistry in microreactors also displays a strategy to overcome classical drawbacks of microwave-driven processes, such as the restricted penetration depth of microwaves into absorbing media.

Green Chemistry Control Keys

There are several controls that can be manipulated at different stages of a chemical manufacturing process:

- Using alternative feedstock or starting materials: Selection of the starting materials has a major effect through the whole synthetic pathway. It determines what hazards will be faced by the workers extracting the substance, shippers transporting the substance, chemists handling the substance. It also predetermines possible future risks from the end-products and wastes. Using more environmentally benign alternative feedstock may improve the environmental profile of the whole process. One of the examples of this step is choosing between the petroleum feedstock and biological feedstock. Currently, 98% of all organic chemicals in the USA are produced from petroleum. Petroleum refining is extremely energy-consuming (15% of total national energy use) and contains high-pollution oxygenation processes. Agricultural feedstocks can be a great alternative that eliminate much of that hazard. Research has shown that many agricultural products (e.g., corn, soy, molasses) can be transformed via variety of processes into textile, nylon, etc.
- Using alternative reagents: Reagents are needed to transform the starting molecules into a target substance. Reagents are not necessarily consumed and are often recycled, but can still bear harm to people and environment exposed to the process. At this point, a chemist must balance the criteria of chemical efficiency and availability with potential hazards.
- Using alternative solvents: Solvents are a very common focusing point because a wide range of syntheses are performed in the liquid media. Many of the currently used solvents are volatile organic compounds. Many of those are responsible for air quality problems (smog, etc.) when released to air. While the traditional organic solvents are easily available, well characterized, and regulated, there is a push for alternative systems that are more environmentally benign in the long run – aqueous solvents, ionic liquids, immobilized solvents, supercritical fluids, etc. The choice of an alternative solvent requires careful and specific analysis, which determines if the new process would be as efficient or as cost-effective.
- Changing target product: Chemistry is function oriented and the target chemical is needed to perform a certain function or possess certain properties. This avenue is related to the search of the alternative final product, which may

require radical change in the way synthesis is done. Through chemical research, it is possible to identify those parts of a molecule that provide the chemical with a desired function as well as those parts that provide toxicity. Maximizing the former and minimizing the latter is a worthy challenge for chemical design.

- Process monitoring: Real time measurements (sensing) of process parameters and concentrations sometimes provide valuable information and hints how the process should be tuned to avoid adverse effects or risk. Also, process monitoring may open avenues for making the process more cost-effective.
- Alternative catalysis: Catalysis bears enormous benefits not only from the standpoint of technical efficiency. Environmental benefit results from the use of a much smaller amount of reagents in catalyzed reactions, which otherwise would contribute to waste stream. Using less chemicals is also economically profitable. It should be noted, though, that many classes of catalysis (e.g., heavy metals) are very toxic. Hence, the challenge of alternative catalysis is to develop environmentally benign options.

Green Chemistry Matrics

Green chemistry metrics are metrics that measure aspects of a chemical process relating to the principles of green chemistry. These metrics serve to quantify the efficiency or environmental performance of chemical processes, and allow changes in performance to be measured. The motivation for using metrics is the expectation that quantifying technical and environmental improvements can make the benefits of new technologies more tangible, perceptible, or understandable. This, in turn, is likely to aid the communication of research and potentially facilitate the wider adoption of green chemistry technologies in industry.

For a non-chemist the most attractive method of quoting the improvement might be a decrease of X unit cost per kilogram of compound Y. This, however, would be an over-simplification—for example, it would not allow a chemist to visualise the improvement made or to understand changes in material toxicity and process hazards. For yield improvements and selectivity increases, simple percentages are suitable, but this simplistic approach may not always be appropriate. For example, when a highly pyrophoric reagent is replaced by a benign one, a numerical value is difficult to assign but the improvement is obvious, if all other factors are similar.

Numerous metrics have been formulated over time and their suitability discussed at great length. A general problem observed is that the more accurate and universally applicable the metric devised, the more complex and unemployable it becomes. A good metric must be clearly defined, simple, measurable, objective rather than subjective and must ultimately drive the desired behavior.

Effective Mass Yield

Effective mass yield is defined as the percentage of the mass of the desired product relative to the mass of all non-benign materials used in its synthesis. Hudlicky *et al.* suggests the following equation:

Effective mass yield (%) = mass of products × 100% / mass of non-benign reagents.

This metric requires further definition of a benign substance. Hudlicky defines it as "those by-products, reagents or solvents that have no environmental risk associated with them, for example, water, low-concentration saline, dilute ethanol, autoclaved cell mass, etc." This definition leaves the metric open to criticism, as nothing is non-benign (which is a subjective term) and the substances listed in the definition have some environmental impact associated with them. The formula also fails to address the level of toxicity associated with a process. Until all toxicology data is available for all chemicals and a term dealing with these levels of "non-benign" reagents is written into the formula the effective mass yield is not the best metric for chemistry.

Carbon Efficiency

Carbon efficiency is how much carbon ends up in the useful product compared to how much carbon was used to create the product.

Carbon efficiency (%) = amount of carbon in product × 100% / total carbon present in reactants.

This metric is a good simplification for use in the pharmaceutical industry as it takes into account the stoichiometry of reactants and products. Furthermore, this metric is of interest to the pharmaceutical industry where development of carbon skeletons is key to their work.

Atom Economy

Atom economy is different from other green chemistry metrics, most of these were designed to measure process improvements. Barry Trost conversely, designed atom economy as a framework by which organic chemists would pursue "greener" chemistry. The atom economy number is how much of the reactants remain in the final product. This is shown below:

For a generic multi-stage reaction:

$$A + B \rightarrow C$$

$$C + D \rightarrow E$$

$$E + F \rightarrow G$$

Atom economy = m.w. of G × 100% / Σ (m.w. A,B,D,F)

The drawback of this type of analysis is that assumptions have to be made. For example, reagents that do not end up in the final product (such as potassium carbonate in a Williamson ether synthesis) are ignored. Also, solvents are ignored. The atom economy calculation is a very simple representation of the "green-ness" of a reaction as it can be carried out without the need for experimental results. However, it can be useful in the process synthesis early stage design.

Reaction Mass Efficiency

The reaction mass efficiency takes into account atom economy, chemical yield and stoichiometry. For a generic reaction A + B → C, the formula can take either of the two forms shown below:

- Reaction mass efficiency = molecular weight of product C × yield / m.w. A + (m.w. B × molar ratio B/A).
- Reaction mass efficiency = mass of product C × 100% / mass of A + mass of B.

The conservation of mass principle dictates that the total molecular mass of the reactants is the same as the total molecular mass of the products. In an ideal chemical process, the amount of starting materials or reactants equals the amount of all products generated and no atom is wasted. However, in some processes, some of the consumed reactant atoms do not become part of the intended products. This can be a concern when raw materials are expensive or when economic and environmental costs of disposal of the waste are high.

Like carbon efficiency, this measure shows the "clean-ness" of a reaction but not of a process. Neither metric takes into account waste produced. For example, these metrics could present a rearrangement as "very green" but fail to address any solvent, work-up, and energy issues that make the process less attractive.

Environmental (E) Factor

The first general metric for green chemistry remains one of the most flexible and popular ones. Roger A. Sheldon's E-factor can be made as complex and thorough or as simple as desired and useful.

The E-factor the ratio of the mass of waste per mass of product:

E-factor = total waste / product

As examples, Sheldon calculated E-factors of various industries.

Table: E-Factors across the chemical industry.

Industry sector	Annual production (t)	E-factor	Waste produced (t)
Oil refining	10^6-10^8	Ca. 0.1	10^5-10^7
Bulk chemicals	10^4-10^6	<1–5	10^4-5×10^6

Fine chemicals	10^2–10^4	5–50	5×10^2–5×10^5
Pharmaceuticals	10–10^3	25–100	2.5×10^2–10^5

It highlights the waste produced in the process as opposed to the reaction, thus helping those who try to fulfil one of the twelve principles of green chemistry to avoid waste production. E-factors ignore recyclable factors such as recycled solvents and re-used catalysts, which obviously increases the accuracy but ignores the energy involved in the recovery (these are often included theoretically by assuming 90% solvent recovery). The main difficulty with E-factors is the need to define system boundaries, for example, which stages of the production or product life-cycle to consider before calculations can be made.

Crucially, this metric is simple to apply industrially, as a production facility can measure how much material enters the site and how much leaves as product and waste, thereby directly giving an accurate global E-factor for the site. Table above shows that oil companies produce a lot less waste than pharmaceuticals as a percentage of material processed. This reflects the fact that the profit margins in the oil industry require them to minimise waste and find uses for products which would normally be discarded as waste. By contrast the pharmaceutical sector is more focused on molecule manufacture and quality. The (currently) high profit margins within the sector mean that there is less concern about the comparatively large amounts of waste that are produced (especially considering the volumes used) although it has to be noted that, despite the percentage waste and E-factor being high, the pharmaceutical section produces much lower tonnage of waste than any other sector. The above table encouraged a number of large pharmaceutical companies to commence "green" chemistry programs.

By incorporating yield, stoichiometry and solvent usage the E-factor is an excellent metric. Crucially, E-factors can be combined to assess multi-step reactions step by step or in one calculation.

The EcoScale

For evaluation of the effectiveness of a synthetic reaction. It is characterized by simplicity and general applicability. Like the yield-based scale, the EcoScale gives a score from 0 to 100, but also takes into account cost, safety, technical set-up, energy and purification aspects. It is obtained by assigning a value of 100 to an ideal reaction defined as "Compound A (substrate) undergoes a reaction with (or in the presence of)inexpensive compound(s) B to give the desired compound C in 100% yield at room temperature with a minimal risk for the operator and a minimal impact on the environment", and then subtracting penalty points for non-ideal conditions. These penalty points take into account both the advantages and disadvantages of specific reagents, set-ups and technologies.

BioLogicTool Plots

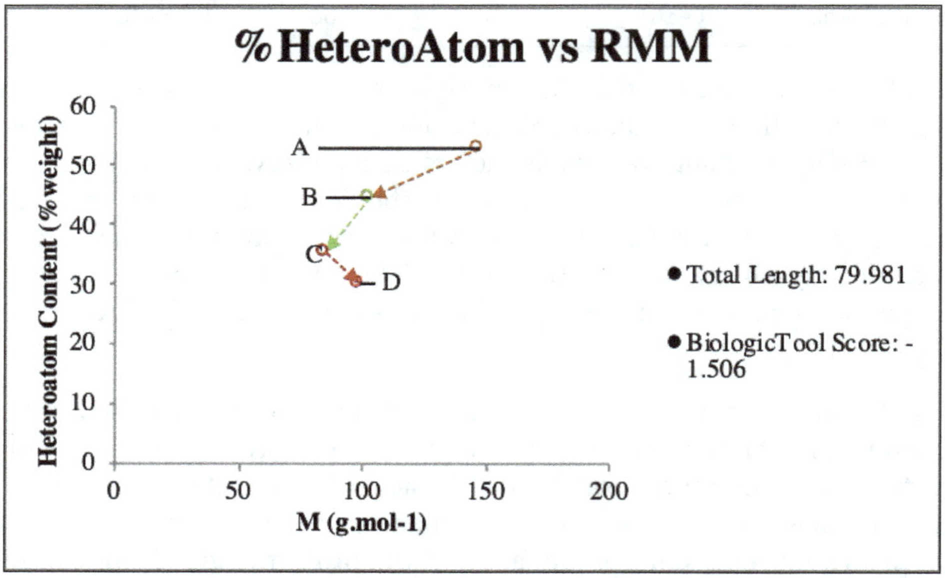

Plots of %heteroatoms against molar mass so called BiologicTool plots.

This freely available tool offers a visual representation of a chemical route from data added by the user (starting material, intermediates and product names, their chemical formula, molar masses and optionally yields of reaction steps). The plots as well as the two scores given were designed to help assessing the rationality of a chemical route with special emphasis on the comparison of bio-based routes with petrol-based ones.

The percentage by mass of heteroatoms contained in the starting material, intermediates, and final product is plotted against their respective molar masses. The visual representation of the chemical routes studied comes with two scores namely, the Total Length and the BioLogictool Score. After normalization of the data, the Total Length is calculated by summation of the individual length of all plotted vectors (commencing from starting material and finishing to intermediates/products). The BioLogicTool score is then obtained after the division of the Total Length by the length of a hypothetical direct vector starting from the original feedstock (A in the example) and finishing to the product (D). The closer to 1 the Biologictool score is and the smaller the Total Length, the more rational the chemical route will be.

Waste Management

Waste management (or waste disposal) include the activities and actions required to manage waste from its inception to its final disposal. This includes the collection, transport, treatment and disposal of waste, together with monitoring and regulation of the waste management process.

Waste can be solid, liquid, or gas and each type has different methods of disposal and management. Waste management deals with all types of waste, including industrial, biological and household. In some cases, waste can pose a threat to human health. Waste is produced by human activity, for example, the extraction and processing of raw materials. Waste management is intended to reduce adverse effects of waste on human health, the environment or aesthetics.

Waste management.

Waste management practices are not uniform among countries (developed and developing nations); regions (urban and rural areas), and residential and industrial sectors can all take different approaches. A large portion of waste management practices deal with municipal solid waste (MSW) which is the bulk of the waste that is created by household, industrial, and commercial activity.

Waste management.

Principles of Waste Management

Waste Hierarchy

The waste hierarchy refers to the "3 Rs" reduce, reuse and recycle, which classifies waste management strategies according to their desirability in terms of waste minimisation. The waste hierarchy is the cornerstone of most waste minimisation strategies. The aim of the waste hierarchy is to extract the maximum practical benefits from products and to generate the minimum amount of end waste. The waste hierarchy is represented as a pyramid because the basic premise is that policies should promote measures to prevent the generation of waste. The next step or preferred action is to seek alternative uses for the waste that has been generated i.e. by re-use. The next is recycling which includes composting. Following this step is material recovery and waste-to-energy. The final action is disposal, in landfills or through incineration without energy recovery. This last step is the final resort for waste which has not been prevented, diverted or recovered. The waste hierarchy represents the progression of a product or material through the sequential stages of the pyramid of waste management. The hierarchy represents the latter parts of the life-cycle for each product.

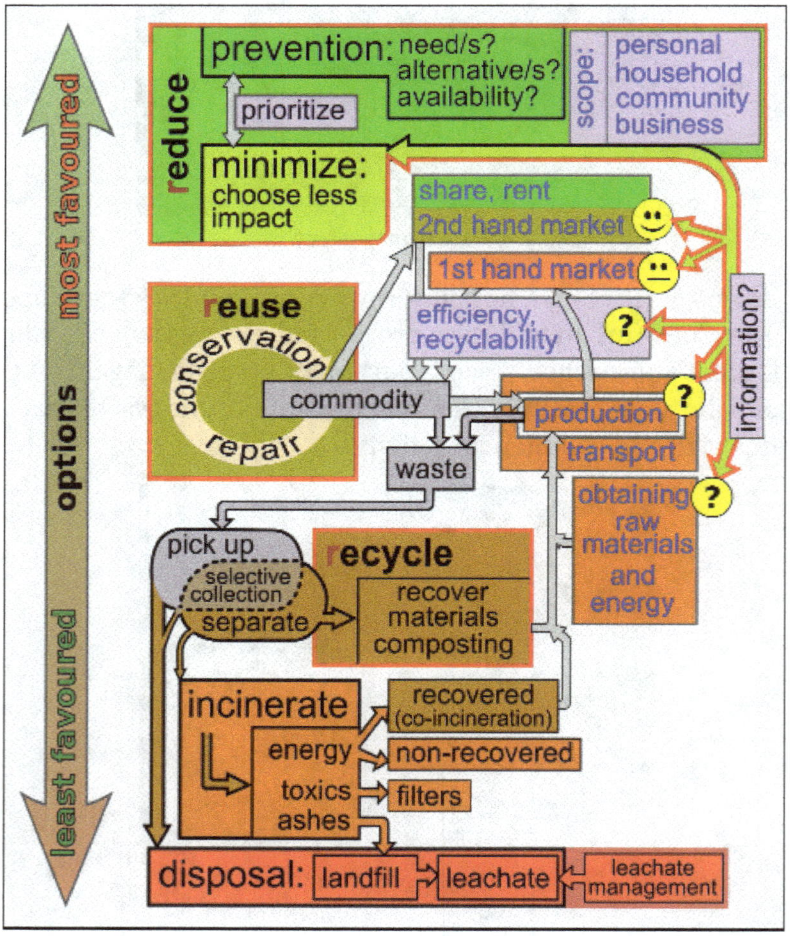

Diagram of the waste hierarchy.

Life-cycle of a Product

The life-cycle begins with design, then proceeds through manufacture, distribution, and primary use and then follows through the waste hierarchy's stages of reduce, reuse and recycle. Each stage in the life-cycle offers opportunities for policy intervention, to rethink the need for the product, to redesign to minimize waste potential, to extend its use. Product life-cycle analysis is a way to optimize the use of the world's limited resources by avoiding the unnecessary generation of waste.

Resource Efficiency

Resource efficiency reflects the understanding that global economic growth and development can not be sustained at current production and consumption patterns. Globally, humanity extracts more resources to produce goods than the planet can replenish. Resource efficiency is the reduction of the environmental impact from the production and consumption of these goods, from final raw material extraction to last use and disposal.

Polluter-pays Principle

The polluter-pays principle mandates that the polluting party pays for the impact on the environment. With respect to waste management, this generally refers to the requirement for a waste generator to pay for appropriate disposal of the unrecoverable material.

Disposal Methods

Landfill

A landfill compaction vehicle in action.

A landfill is a site for the disposal of waste materials by burial. Landfill is the oldest form of waste treatment, although the burial of the waste is modern; historically, refuse was simply left in piles or thrown into pits. Historically, landfills have been the most common method of organized waste disposal and remain so in many places around the world.

Incineration

Incineration is a disposal method in which solid organic wastes are subjected to combustion so as to convert them into residue and gaseous products. This method is useful for disposal of both municipal solid waste and solid residue from waste water treatment. This process reduces the volumes of solid waste by 80 to 95 percent. Incineration and other high temperature waste treatment systems are sometimes described as "thermal treatment". Incinerators convert waste materials into heat, gas, steam, and ash.

Spittelau incineration plant.

Incineration is carried out both on a small scale by individuals and on a large scale by industry. It is used to dispose of solid, liquid and gaseous waste. It is recognized as a practical method of disposing of certain hazardous waste materials (such as biological medical waste). Incineration is a controversial method of waste disposal, due to issues such as emission of gaseous pollutants.

Incineration is common in countries such as Japan where land is more scarce, as the facilities generally do not require as much area as landfills. Waste-to-energy (WtE) or energy-from-waste (EfW) are broad terms for facilities that burn waste in a furnace or boiler to generate heat, steam or electricity. Combustion in an incinerator is not always perfect and there have been concerns about pollutants in gaseous emissions from incinerator stacks. Particular concern has focused on some very persistent organic compounds such as dioxins, furans, and PAHs, which may be created and which may have serious environmental consequences.

Recycling

Recycling is a resource recovery practice that refers to the collection and reuse of waste materials such as empty beverage containers. The materials from which the items are made can be reprocessed into new products. Material for recycling may be collected separately from general waste using dedicated bins and collection vehicles, a procedure called kerbside collection. In some communities, the owner of the waste is required to separate the materials into different bins (e.g. for paper, plastics, metals) prior to its collection. In other communities, all recyclable materials are placed in a single bin for collection, and the sorting is handled later at a central facility. The latter method is known as "single-stream recycling."

Waste not the Waste.

The most common consumer products recycled include aluminium such as beverage cans, copper such as wire, steel from food and aerosol cans, old steel furnishings or equipment, rubber tyres, polyethylene and PET bottles, glass bottles and jars, paperboard cartons, newspapers, magazines and light paper, and corrugated fiberboard boxes. PVC, LDPE, PP, and PS are also recyclable. These items are usually composed of a single type of material, making them relatively easy to recycle into new products. The recycling of complex products (such as computers and electronic equipment) is more difficult, due to the additional dismantling and separation required.

Steel crushed and baled for recycling.

The type of material accepted for recycling varies by city and country. Each city and country has different recycling programs in place that can handle the various types of recyclable materials. However, certain variation in acceptance is reflected in the resale value of the material once it is reprocessed.

Re-use

Biological Reprocessing

Recoverable materials that are organic in nature, such as plant material, food scraps, and paper products, can be recovered through composting and digestion processes to decompose the organic matter. The resulting organic material is then recycled as mulch or compost for agricultural or landscaping purposes. In addition, waste gas from the process (such as methane) can be captured and used for generating electricity and heat (CHP/cogeneration) maximising efficiencies. The intention of biological processing in waste management is to control and accelerate the natural process of decomposition of organic matter.

An active compost heap.

Energy Recovery

Energy recovery from waste is the conversion of non-recyclable waste materials into usable heat, electricity, or fuel through a variety of processes, including combustion, gasification, pyrolyzation, anaerobic digestion, and landfill gas recovery. This process is often called waste-to-energy. Energy recovery from waste is part of the non-hazardous waste management hierarchy. Using energy recovery to convert non-recyclable waste materials into electricity and heat, generates a renewable energy source and can reduce carbon emissions by offsetting the need for energy from fossil sources as well as reduce methane generation from landfills. Globally, waste-to-energy accounts for 16% of waste management.

The energy content of waste products can be harnessed directly by using them as a direct combustion fuel, or indirectly by processing them into another type of fuel. Thermal treatment ranges from using waste as a fuel source for cooking or heating and the use of the gas fuel, to fuel for boilers to generate steam and electricity in a turbine. Pyrolysis and gasification are two related forms of thermal treatment where waste materials are heated to high temperatures with limited oxygen availability. The process usually occurs in a sealed vessel under high pressure. Pyrolysis of solid waste converts the material into solid, liquid and gas products. The liquid and gas can be burnt to produce energy or refined into other chemical products (chemical refinery). The solid residue (char) can be further refined into products such as activated carbon. Gasification and advanced Plasma arc gasification are used to convert organic materials directly into a synthetic gas (syngas) composed of carbon monoxide and hydrogen. The gas is then burnt to produce electricity and steam. An alternative to pyrolysis is high temperature and pressure supercritical water decomposition (hydrothermal monophasic oxidation).

Pyrolysis

Pyrolysis is often used to convert many types of domestic and industrial residues into a recovered fuel. Different types of waste input (such as plant waste, food waste, tyres) placed in the pyrolysis process potentially yield an alternative to fossil fuels. Pyrolysis is a process of thermo-chemical decomposition of organic materials by heat in the absence of stoichiometric quantities of oxygen; the decomposition produces various hydrocarbon gases. During pyrolysis, the molecules of object vibrate at high frequencies to an extent that molecules start breaking down. The rate of pyrolysis increases with temperature. In industrial applications, temperatures are above 430 °C (800 °F). Slow pyrolysis produces gases and solid charcoal. Pyrolysis hold promise for conversion of waste biomass into useful liquid fuel. Pyrolysis of waste wood and plastics can potentially produce fuel. The solids left from pyrolysis contain metals, glass, sand and pyrolysis coke which does not convert to gas. Compared to the process of incineration, certain types of pyrolysis processes release less harmful by-products that contain alkali metals, sulphur, and chlorine. However, pyrolysis of some waste yields gasses which impact the environment such as HCl and SO_2.

Resource Recovery

Resource recovery is the systematic diversion of waste, which was intended for disposal, for a specific next use. It is the processing of recyclables to extract or recover materials and resources, or convert to energy. These activities are performed at a resource recovery facility. Resource recovery is not only environmentally important, but it is also cost-effective. It decreases the amount of waste for disposal, saves space in landfills, and conserves natural resources.

Resource recovery (as opposed to waste management) uses LCA (life cycle analysis) attempts to offer alternatives to waste management. For mixed MSW (Municipal Solid Waste) a number of broad studies have indicated that administration, source separation and collection followed by reuse and recycling of the non-organic fraction and energy and compost/fertilizer production of the organic material via anaerobic digestion to be the favoured path.

As an example of how resource recycling can be beneficial, many items thrown away contain metals that can be recycled to create a profit, such as the components in circuit boards. Wood chippings in pallets and other packaging materials can be recycled to useful products for horticulture. The recycled chips can cover paths, walkways, or arena surfaces.

Application of rational and consistent waste management practices can yield a range of benefits including:

- Economic: Improving economic efficiency through the means of resource use, treatment and disposal and creating markets for recycles can lead to efficient practices in the production and consumption of products and materials resulting in valuable materials being recovered for reuse and the potential for new jobs and new business opportunities.
- Social: By reducing adverse impacts on health by proper waste management practises, the resulting consequences are more appealing civic communities. Better social advantages can lead to new sources of employment and potentially lifting communities out of poverty especially in some of the developing poorer countries and cities.
- Environmental: Reducing or eliminating adverse impacts on the environment through reducing, reusing and recycling, and minimizing resource extraction can result in improved air and water quality and help in the reduction of greenhouse gas emissions.
- Inter-generational Equity: Following effective waste management practises can provide subsequent generations a more robust economy, a fairer and more inclusive society and a cleaner environment.

Sustainability

The management of waste is a key component in a business' ability to maintain ISO14001 accreditation. The standard encourages companies to improve their environmental efficiencies each year by eliminating waste through resource recovery practices. One way to do this is by adopting resource recovery practices like recycling materials such as glass, food scraps, paper and cardboard, plastic bottles and metal. Recycled materials can often be sold to the construction industry. Many inorganic waste streams can be used to produce materials for construction. Concrete and bricks can be recycled as artificial gravel.

Liquid Waste-management

- Sewage Sludge: Sewage sludge is produced by waste water treatment processes. Due to rapid urbanization, there has been an increase in municipal waste water that results 0.1–30.8 kg of sewage per population equivalent per year (kg/p.e/year). Common disposal practices of sewage sludge are incineration, composting, and landfill.
- Avoidance and Reduction Methods: An important method of waste management is the prevention of waste material being created, also known as waste reduction. Methods of avoidance include reuse of second-hand products, repairing

broken items instead of buying new ones, designing products to be refillable or reusable (such as cotton instead of plastic shopping bags), encouraging consumers to avoid using disposable products (such as disposable cutlery), removing any food/liquid remains from cans and packaging, and designing products that use less material to achieve the same purpose (for example, lightweighting of beverage cans).

- Technologies: Traditionally, the waste management industry has been a late adopter of new technologies such as RFID (Radio Frequency Identification) tags, GPS and integrated software packages which enable better quality data to be collected without the use of estimation or manual data entry.

Waste Minimisation

Waste minimisation is a set of processes and practices intended to reduce the amount of waste produced. By reducing or eliminating the generation of harmful and persistent wastes, waste minimisation supports efforts to promote a more sustainable society. Waste minimisation involves redesigning products and processes and/or changing societal patterns of consumption and production.

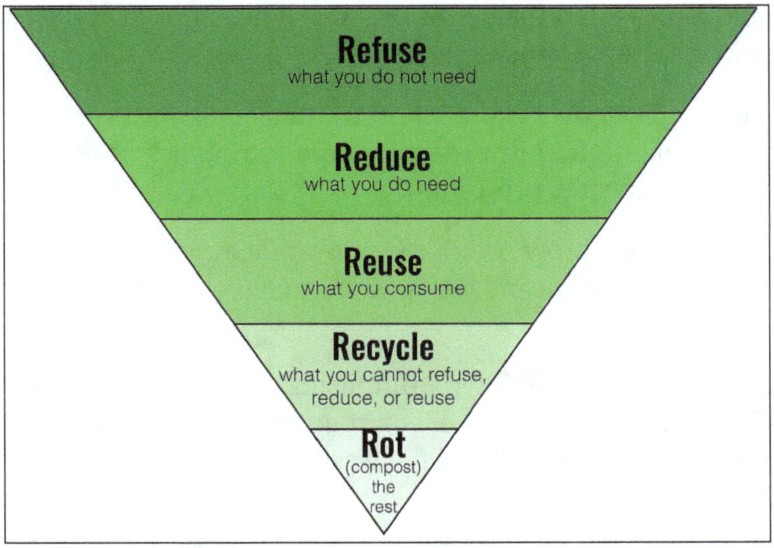

Waste hierarchy: Refusing, reducing, reusing, recycling and composting allow to reduce waste.

The most environmentally resourceful, economically efficient, and cost effective way to manage waste often is to not have to address the problem in the first place. Managers see waste minimisation as a primary focus for most waste management strategies. Proper waste treatment and disposal can require a significant amount of time and resources; therefore, the benefits of waste minimisation can be considerable if carried out in an effective, safe and sustainable manner.

Traditional waste management focuses on processing waste after it is created, concentrating on re-use, recycling, and waste-to-energy conversion. Waste minimisation involves efforts to avoid creating the waste during manufacturing. To effectively implement waste minimisation the manager requires knowledge of the production process, cradle-to-grave analysis (the tracking of materials from their extraction to their return to earth) and details of the composition of the waste.

The main sources of waste vary from country to country. Most waste comes from the construction and demolition of buildings, followed by mining and quarrying, industry and commerce. Household waste constitutes a relatively small proportion of all waste. Industrial waste is often tied to requirements in the supply chain. For example, a company handling a product may insist that it should be shipped using particular packing because it fits downstream needs.

Benefits of Waste Minimisation

Waste minimisation can protect the environment and often turns out to have positive economic benefits. Waste minimisation can improve:

- Efficient production practices: Waste minimisation can achieve more output of product per unit of input of raw materials.
- Economic returns: More efficient use of products means reduced costs of purchasing new materials improving the financial performance of a company.
- Public image: The environmental profile of a company is an important part of its overall reputation and waste minimisation reflects a proactive movement towards environmental protection.
- Quality of products produced: New innovation and technological practices can reduce waste generation and improve the quality of the inputs in the production phase.
- Environmental responsibility: Minimising or eliminating waste generation makes it easier to meet targets of environmental regulations, policies, and standards. The environmental impact of waste will be reduced.

Industries

In industry, using more efficient manufacturing processes and better materials generally reduces the production of waste. The application of waste minimisation techniques has led to the development of innovative and commercially successful replacement products. Waste minimisation efforts often require investment, which is usually compensated by the savings. However, waste reduction in one part of the production process may create waste production in another part.

Processes

- Reuse of scrap material: Scraps can be immediately re-incorporated at the beginning of the manufacturing line so that they do not become a waste product. Many industries routinely do this; for example, paper mills return any damaged rolls to the beginning of the production line, and in the manufacture of plastic items, off-cuts and scrap are re-incorporated into new products.
- Improved quality control and process monitoring: Steps can be taken to ensure that the number of reject batches is kept to a minimum. This is achieved by increasing the frequency of inspection and the number of points of inspection. For example, installing automated continuous monitoring equipment can help to identify production problems at an early stage.
- Waste exchanges: This is where the waste product of one process becomes the raw material for a second process. Waste exchanges represent another way of reducing waste disposal volumes for waste that cannot be eliminated.
- Ship to point of use: This involves making deliveries of incoming raw materials or components direct to the point where they are assembled or used in the manufacturing process to minimise handling and the use of protective wrappings or enclosures.
- Zero waste: This is a whole systems approach that aims to eliminate waste at the source and at all points down the supply chain, with the intention of producing no waste. It is a design philosophy which emphasizes waste prevention as opposed to end of pipe waste management. Since, globally speaking, waste as such, however minimal, can never be prevented (there will always be an end-of-life even for recycled products and materials), a related goal is pollution prevention.

Product Design

Waste minimisation and resource maximisation for manufactured products can most easily be done at the design stage. Reducing the number of components used in a product or making the product easier to take apart can make it easier to be repaired or recycled at the end of its useful life. In some cases, it may be best not to minimise the volume of raw materials used to make a product, but instead reduce the volume or toxicity of the waste created at the end of a product's life, or the environmental impact of the product's use.

Fitting the Intended Use

In this strategy, products and packages are optimally designed to meet their intended use. This applies especially to packaging materials, which should only be as durable as necessary to serve their intended purpose. On the other hand, it could be more wasteful if food, which has consumed resources and energy in its production, is damaged and spoiled because of extreme measures to reduce the use of paper, metals, glass and plastics in its packaging.

Durability

Improving product durability, such as extending a vacuum cleaner's useful life to 15 years instead of 12, can reduce waste and usually much improves resource optimisation. But in some cases it has a negative environmental impact. If a product is too durable, its replacement with more efficient technology is likely to be delayed. Therefore, extending an older machine's useful life may place a heavier burden on the environment than scrapping it, recycling its metal and buying a new model. Similarly, older vehicles consume more fuel and produce more emissions than their modern counterparts. Most proponents of waste minimisation consider that the way forward may be to view any manufactured product at the end of its useful life as a resource for recycling and reuse rather than waste.

Making refillable glass bottles strong enough to withstand several journeys between the consumer and the bottling plant requires making them thicker and so heavier, which increases the resources required to transport them. Since transport has a large environmental impact, careful evaluation is required of the number of return journeys bottles make. If a refillable bottle is thrown away after being refilled only several times, the resources wasted may be greater than if the bottle had been designed for a single journey. Many choices involve trade-offs of environmental impact, and often there is insufficient information to make informed decisions.

Atom Economy

Atom economy (atom efficiency/percentage) is the conversion efficiency of a chemical process in terms of all atoms involved and the desired products produced. Atom economy is an important concept of green chemistry philosophy, and one of the most widely used metrics for measuring the "greenness" of a process or synthesis.

Atom economy can be written as:

$$\text{Atom economy} = \frac{\text{Molecular Weight of desired product}}{\text{Molecular Weight of all reactants}} \times 100\%$$

For a multi-step process, where the intermediates are formed in one step and consumed during a later step:

$$A + B \rightarrow C$$
$$C + D \rightarrow E$$
$$E + F \rightarrow G$$

$$\text{Atom economy} = \frac{M_r(G)}{M_r(A)+M_r(B)+M_r(D)+M_r(F)}$$

Atom economy is a different concern than chemical yield, because a high-yielding process can still result in substantial byproducts. Examples include the Cannizzaro reaction, in which approximately 50% of the reactant aldehyde becomes the other oxidation state of the target; the Wittig reaction, which uses high-mass phosphorus reagents that ultimately become waste; and the Gabriel synthesis, which produces a stoichiometric quantity of phthalic acid.

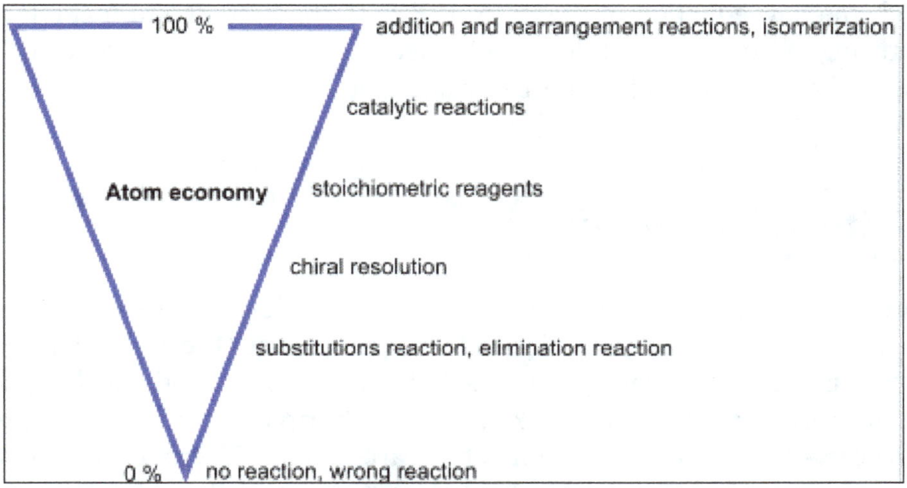

Atom economy.

If the desired product has an enantiomer the reaction needs to be sufficiently stereoselective even when atom economy is 100%. A Diels-Alder reaction is an example of a potentially very atom efficient reaction that also can be chemo-, regio-, diastereo- and enantioselective. Catalytic hydrogenation comes the closest to being an ideal reaction that is extensively practiced both industrially and academically.

Atom economy can also be adjusted if a pendant group is recoverable, for example Evans auxiliary groups. However, if this can be avoided it is more desirable, as recovery processes will never be 100%. Atom economy can be improved upon by careful selection of starting materials and a catalyst system.

Poor atom economy is common in fine chemicals or pharmaceuticals synthesis, and especially in research, where the aim to readily and reliably produce a wide range of complex compounds leads to the use of versatile and dependable, but poorly atom-economical reactions. For example, synthesis of an alcohol is readily accomplished by reduction of an ester with lithium aluminium hydride, but the reaction necessarily produces a voluminous floc of aluminum salts, which have to be separated from the product alcohol and disposed of. The cost of such hazardous material disposal can be considerable. Catalytic hydrogenolysis of an ester is the analogous reaction with a high atom economy, but it requires catalyst optimization, is a much slower reaction and is not applicable universally.

Creating Reactions Utilizing Atom Economy

It is fundamental in chemical reactions of the form A+B → C+D that two products are necessarily generated though product C may have been the desired one. That being the case, D is considered a byproduct. As it is a significant goal of green chemistry to maximize the efficiency of the reactants and minimize the production of waste, D must either be found to have use, be eliminated or be as insignificant and innocuous as possible. With the new equation of the form A+B → C, the first step in making chemical manufacturing more efficient is the use of reactions that resemble simple addition reactions with the only other additions being catalytic materials.

Addition Reaction

An addition reaction, in organic chemistry, is in its simplest terms an organic reaction where two or more molecules combine to form a larger one (the adduct). Addition reactions are limited to chemical compounds that have multiple bonds, such as molecules with carbon–carbon double bonds (alkenes), or with triple bonds (alkynes). Molecules containing carbon—hetero double bonds like carbonyl (C=O) groups, or imine (C=N) groups, can undergo addition, as they too have double-bond character.

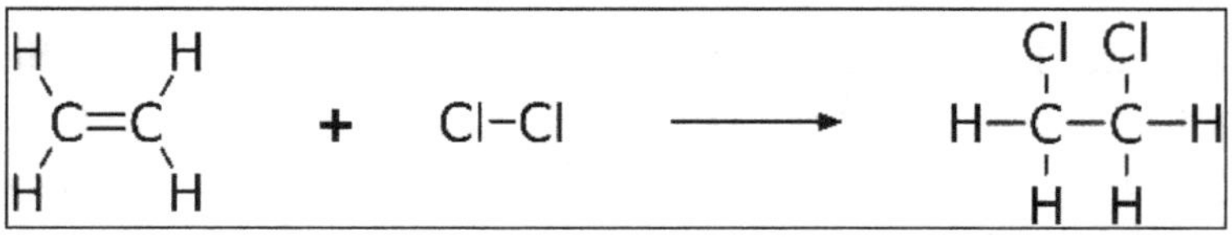

Addition of chlorine to ethylene.

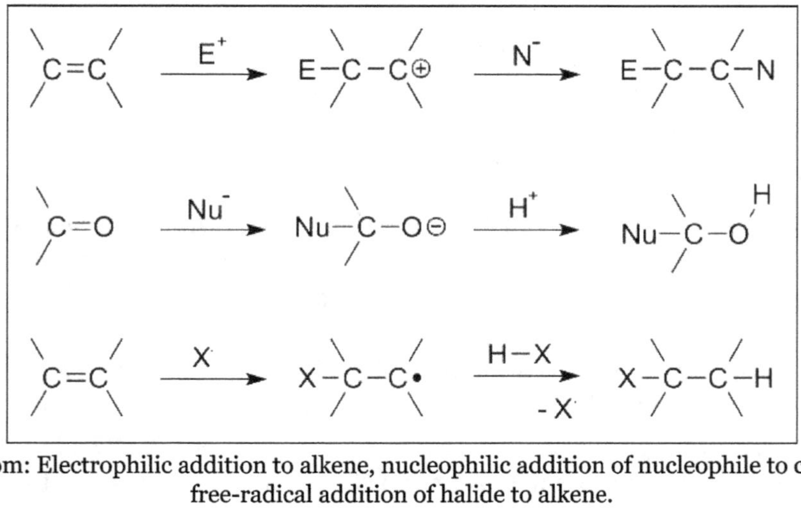

Top to bottom: Electrophilic addition to alkene, nucleophilic addition of nucleophile to carbonyl and free-radical addition of halide to alkene.

An addition reaction is the reverse of an elimination reaction. For instance, the hydration of an alkene to an alcohol is reversed by dehydration. There are two main types of polar addition reactions: electrophilic addition and nucleophilic addition. Two non-polar addition reactions exist as well, called free-radical addition and cycloadditions. Addition reactions are also encountered in polymerizations and called addition polymerization. Depending on the product structure, it could promptly react further to eject a leaving group to give the addition–elimination reaction sequence.

Substitution Reaction

Substitution reaction (also known as single displacement reaction or single substitution reaction) is a chemical reaction during which one functional group in a chemical compound is replaced by another functional group. Substitution reactions are of prime importance in organic chemistry. Substitution reactions in organic chemistry are classified either as electrophilic or nucleophilic depending upon the reagent involved.

Organic substitution reactions are classified in several main organic reaction types depending on whether the reagent that brings about the substitution is considered an electrophile or a nucleophile, whether a reactive intermediate involved in the reaction is a carbocation, a carbanion or a free radical or whether the substrate is aliphatic or aromatic. Detailed understanding of a reaction type helps to predict the product outcome in a reaction. It also is helpful for optimizing a reaction with regard to variables such as temperature and choice of solvent.

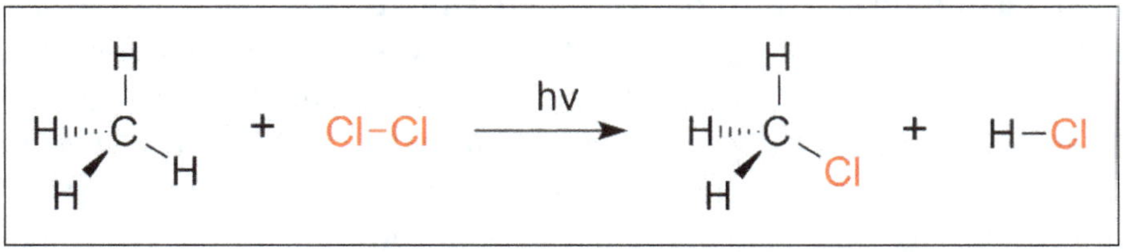

Chlorination of methane by chlorine.

A good example of a substitution reaction is halogenation. When chlorine gas (Cl_2) is irradiated, some of the molecules are split into two chlorine radicals (Cl•) whose free electrons are strongly nucleophilic. One of them breaks a C–H covalent bond in CH_4 and grabs the hydrogen atom to form the electrically neutral HCl. The other radical reforms a covalent bond with the CH_3• to form CH_3Cl (methyl chloride).

Nucleophilic Substitution

In organic (and inorganic) chemistry, nucleophilic substitution is a fundamental class of reactions in which a nucleophile selectively bonds with or attacks the positive or partially positive charge on an atom or a group of atoms. As it does so, it replaces a weaker nucleophile which then becomes a leaving group. The remaining positive or partially positive atom becomes an electrophile. The whole molecular entity of which the electrophile and the leaving group are part is usually called the substrate.

The most general form for the reaction may be given as where R-LG indicates the substrate:

Nuc: + R-LG → R-Nuc + LG

The electron pair (:) from the nucleophile (Nuc:) attacks the substrate (R-LG) forming a new covalent bond Nuc-R-LG. The prior state of charge is restored when the leaving group (LG) departs with an electron pair. The principal product in this case is R-Nuc. In such reactions, the nucleophile is usually electrically neutral or negatively charged, whereas the substrate is typically neutral or positively charged.

An example of nucleophilic substitution is the hydrolysis of an alkyl bromide, R-Br, under basic conditions, where the *attacking* nucleophile is the base OH^- and the leaving group is Br^-:

$$R\text{-}Br + OH^- \rightarrow R\text{-}OH + Br^-$$

Nucleophilic substitution reactions are commonplace in organic chemistry, and they can be broadly categorized as taking place at a carbon of a saturated aliphatic compound carbon or (less often) at an aromatic or other unsaturated carbon center.

Mechanisms

These substitutions can be produced by two different mechanisms categorized at: unimolecular nucleophilic substitution (S_N1) and bimolecular nucleophilic substitution (S_N2). The S_N1 mechanism has two steps. In the first step, the leaving group departs, forming a carbocation C^+. In the second step, the nucleophilic reagent (Nuc:) attaches to the carbocation and forms a covalent sigma bond. If the substrate has a chiral carbon, this mechanism can result in either inversion of the stereochemistry or retention of configuration. Usually both occur without preference. The result is racemization.

The S_N2 mechanism has just one step. The attack of the reagent and the expulsion of the leaving group happen simultaneously. This mechanism always results in inversion of configuration. If the substrate that is under nucleophilic attack is chiral, the reaction will lead to an inversion of its stereochemistry called a Walden inversion.

S_N2 attack may occur if the backside route of attack is not sterically hindered by substituents on the substrate. Therefore, this mechanism usually occurs at an unhindered primary carbon center. If there is steric crowding on the substrate near the leaving group, such as at a tertiary carbon center, the substitution will involve an S_N1 rather than an S_N2 mechanism, (an S_N1 would also be more likely in this case because a sufficiently stable carbocation intermediary could be formed).

When the substrate is an aromatic compound the reaction type is nucleophilic aromatic substitution. Carboxylic acid derivatives react with nucleophiles in nucleophilic acyl substitution. This kind of reaction can be useful in preparing compounds.

Electrophilic Substitution

Electrophiles are involved in electrophilic substitution reactions, particularly in electrophilic aromatic substitutions. In this example the benzene ring's electron resonance structure is attacked by an electrophile E^+. The resonating bond is broken and a carbocation resonating structure results. Finally a proton is kicked out and a new aromatic compound is formed. Electrophilic reactions to other unsaturated compounds than arenes generally lead to electrophilic addition rather than substitution.

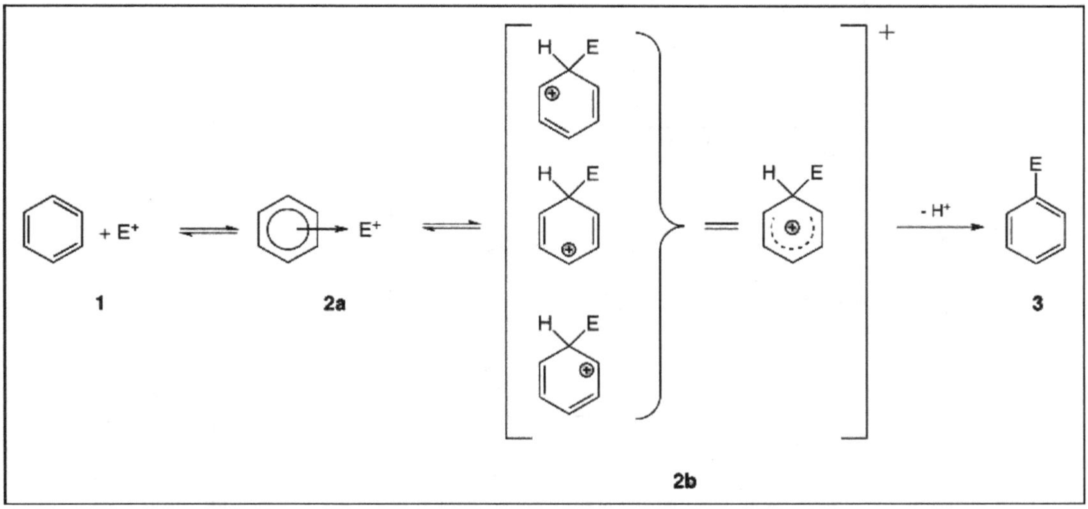

Electrophilic aromatic substitution.

Radical Substitution

A radical substitution reaction involves radicals. An example is the Hunsdiecker reaction.

Organometallic Substitution

Coupling reactions are a class of metal-catalyzed reactions involving an organometallic compound RM and an organic halide R′X that together react to form a compound of the type R-R′ with formation of a new carbon–carbon bond. Examples include the Heck reaction, Ullmann reaction, and Wurtz–Fittig reaction. Many variations exist.

Substituted Compounds

Substituted compounds are chemical compounds where one or more hydrogen atoms of a core structure have been replaced with a functional group like alkyl, hydroxy, or halogen, or with larger substituent groups. For example, benzene is a simple aromatic ring. Benzenes that have undergone substitution are a heterogeneous group of chemicals with a wide spectrum of uses and properties.

Table: Examples of substituted benzene compounds.

Compound	General formula	General Structure
Benzene	C_6H_6	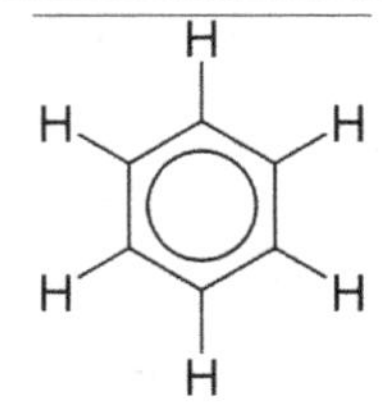
Toluene	$C_6H_5\text{-}CH_3$	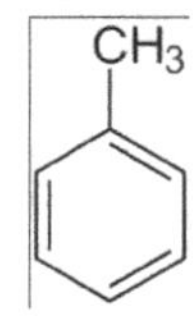
o-Xylene	$C_6H_4(\text{-}CH_3)_2$	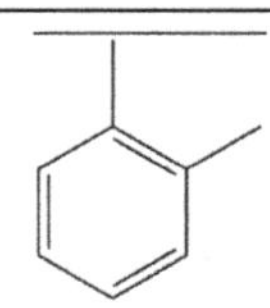
Mesitylene	$C_6H_3(\text{-}CH_3)_3$	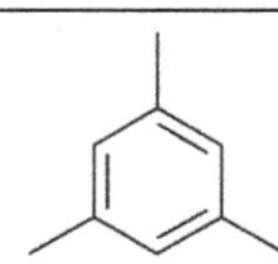
Phenol	$C_6H_5\text{-}OH$	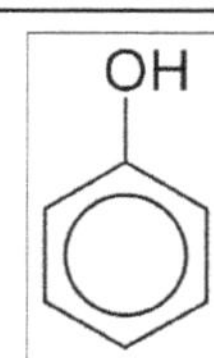

Elimination Reaction

Elimination reaction often competes with substitution reactions. In this reaction, a substrate (typically an alkyl halide) eliminates one equivalent (unit) of acid to form an alkene. Two possible mechanisms are available for this elimination reaction – E1 and E2 mechanisms.

Elimination reaction is a type of reaction is mainly used to transform saturated compounds (organic compounds which contain single carbon-carbon bonds) to unsaturated compounds (compounds which feature double or triple carbon-carbon bonds).

Besides, it is an important method for the preparation of alkenes.

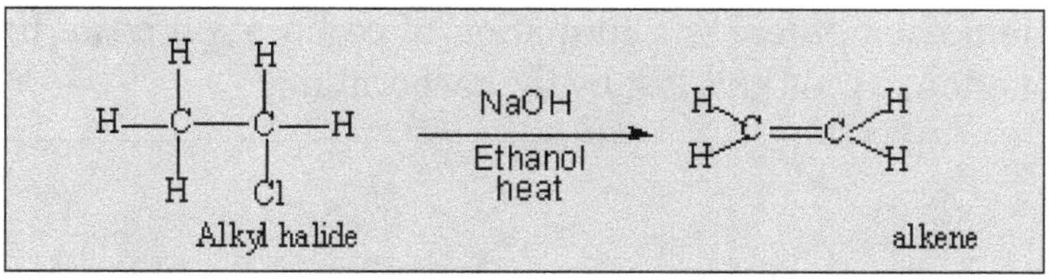

Preparation of Alkenes.

An elimination reaction is a type of a chemical reaction where several atoms either in pairs or groups are removed from a molecule. The removal usually takes place due to the action of acids and bases or action of metals. It can also happen through the process of heating at high temperatures.

Important Methods of Elimination Reaction

Normally, elimination reactions are distinguished by the kind of atoms or groups of atoms that leave the molecule. Due to this, there are two main methods involved in this type of reaction:

- Dehydration.
- Dehydrohalogenation.

In the dehydration method, there is the elimination of a water molecule mostly from compounds such as alcohol. Sometimes, this method is also called Beta elimination reaction where the leaving group and H are placed at neighbour carbon atoms. On the other hand, in dehydrohalogenation, there is a removal of a hydrogen atom and a halogen atom. Some of the other common types of elimination reactions are α-elimination and γ- and δ-elimination.

Mechanism of Elimination Reaction

The elimination reaction consists of three fundamental events and they are:

- Proton removal.
- C-C pi bond is formed.
- There is a breakage in the bond of the leaving group.

Depending on the reaction kinetics, elimination reactions can occur mostly by two mechanisms namely E1 or E2 where E is referred to as elimination and the number represent the molecularity.

E1 Reaction

- In the E1 mechanism which is also known as unimolecular elimination, there are usually two steps involved – ionization and deprotonation.

- During ionization, there is a formation of carbocation as an intermediate. In deprotonation, a proton is lost by the carbocation.

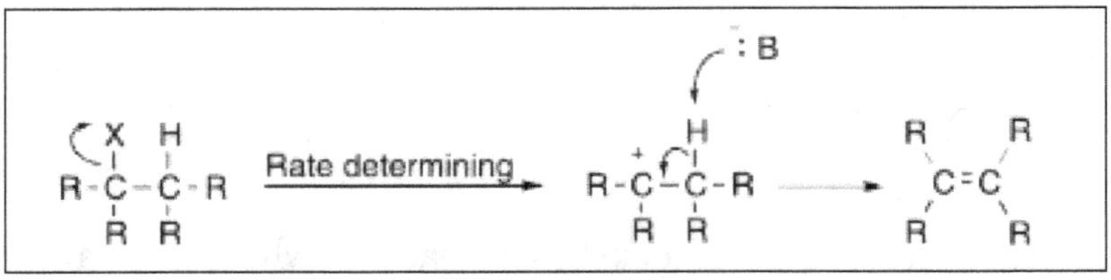

E1 Reaction Mechanism.

- This happens in the presence of a base which further leads to the formation of a pi-bond in the molecule.
- In E1, the reaction rate is also proportional to the concentration of the substance to be transformed.
- It exhibits first-order kinetics.

E 1 mechanism shares the features of the SN1 reaction. The initial step is the formation of a carbocation intermediate through the loss of the leaving group. This slow step becomes the rate-determining step for the whole reaction.

E2 Reaction

- In an E2 mechanism which refers to bimolecular elimination is basically a one-step mechanism.
- Here, the carbon-hydrogen and carbon-halogen bonds mostly break off to form a new double bond.

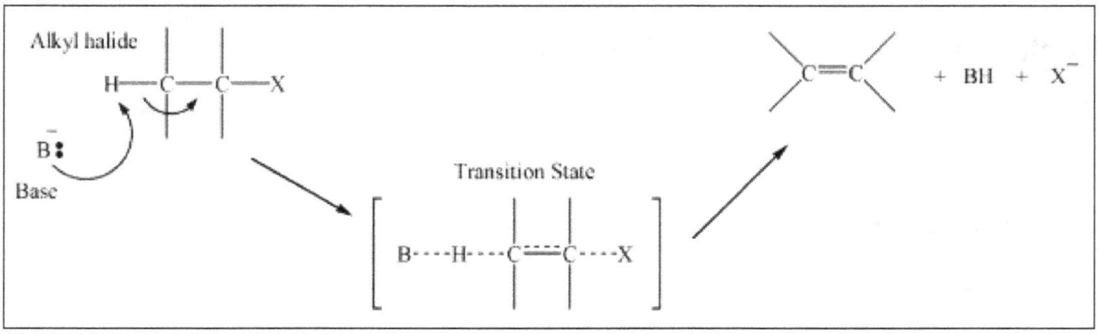

E2 Reaction Mechanism.

- However, in the E2 mechanism, a base is part of the rate determining step and it has a huge influence on the mechanism.
- The reaction rate is mostly proportional to the concentrations of both the eliminating agent and the substrate.
- It exhibits second-order kinetics.

The E2 mechanism can generally be represented as below. B stands for base and X stands for the halogen.

The rate of the E2 reaction is,

Rate = k[RX][Base]

So the reaction rate depends on both the substrate (RX) and the base involved. In the elimination reaction, the major product formed is the most stable alkene.

Elimination Reaction Example

One of the common examples of elimination reaction is the dehydration of alcohol. Here the process takes place in the presence of a base such as an ethoxide ion ($C_2H_5O^-$). It can be represented as;

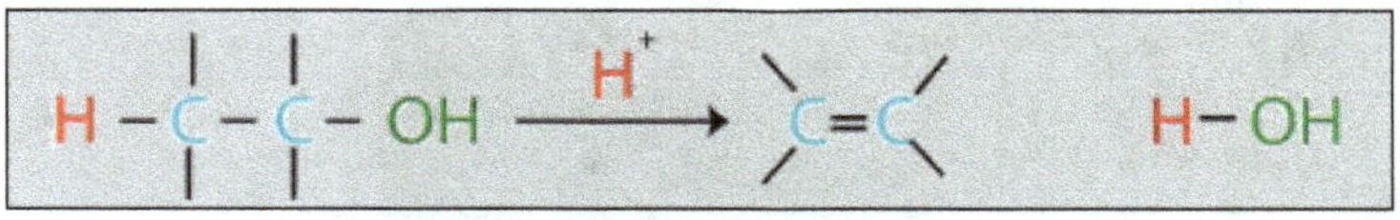

Elimination Reaction Mechanism.

"E2 and E1 reactions differ significantly in the nature of the transition states that determine the regiochemistry of the product". The E2 pathway involves a transition state leading from starting material directly to the product. The product forming step of an E1 reaction is more exothermic than that of an E2 reaction. Thus the E1 reaction has a relatively early transition state, closely resembling the carbocation formed in the rate-determining step.

References

- E. Vedejs & C. F. Marth (1988). "Mechanism of the Wittig reaction: the role of substituents at phosphorus". J. Am. Chem. Soc. 110 (12): 3948–3958. doi:10.1021/ja00220a037
- Hamlin, Trevor A.; van Beek, Bas; Wolters, Lando P.; Bickelhaupt, F. Matthias (2018). "Nucleophilic Substitution in Solution: Activation Strain Analysis of Weak and Strong Solvent Effects". Chemistry – A European Journal. 24 (22): 5927–5938
- Introductory-chapter-principles-of-green-chemistry, green-chemistry: .intechopen.com, Retrieved 11 August, 2020
- Sheldon R. A. (2000). "Atom efficiency and catalysis in organic synthesis" (PDF). Pure and Applied Chemistry. 72 (7): 1233–1246. doi:10.1351/pac200072071233
- Guidelines for National Waste Management Strategies Moving from Challenges to Opportunities (PDF). United Nations Environmental Programme. 2013. ISBN 978-92-807-3333-4

Green Synthesis and Chemistry

3

- Green Synthesis of Oligomer Calixarenes
- Green Chemistry in Organic Synthesis
- Synthesis and Characterization of Nanocomplexes

Green synthesis provides environmental and economic benefits through the use of non-toxic and safe reagents. It avoids the production of unwanted or harmful by-products through the build-up of reliable, sustainable, and eco-friendly synthesis procedures. This chapter discusses the subject of green synthesis and chemistry in detail.

Green Synthesis of Oligomer Calixarenes

A calixarene is a cup-shaped supramolecule. The name is derived from the word "calix," which means a cup. Synthesis of calixarenes becomes very important because these molecules are widely used in various fields, especially in its use in a guest-host system molecule. The molecule was first reported in 1872 by Baeyer as a reaction product of aldehyde-phenol condensation. The same product can also be formed from benzaldehyde-pyrogallol (benzene-1,2,3-triol) condensation reaction. The structure of phenols used in this reaction has reactivity at ortho and para positions so that it can form a polymer cross-linked at three positions (two orthos and one para). In 1942, Zinke and Ziegler used para-substituted phenols so that the phenols could only react at the two ortho positions, thereby reducing the possibility of cross-link formation. By reacting *p-tert*-butylphenol with formaldehyde under a basic condition at 100–220°C, a waxy brown paste was produced, which was then recrystallized and formed a crystal decomposed at above 300°C. The results of the analysis showed that the product had a molecular weight that was in agreement with a cyclic tetrameric structure. Similarly, Cornforth reacted *p-tert*-butylphenol with formaldehyde. However, the reaction resulted in two crystal products that were rather soluble in water with high but nonidentical melting points. Likewise, when *p-tert*-butylphenol was replaced with *p*-1,1,3,3-tetramethylbutylphenol, two products with different melting points were obtained. The results of X-ray showed that the four condensation products had a cyclic tetrameric structure. The different melting points,indicated that the two compounds are diastereoisomers.

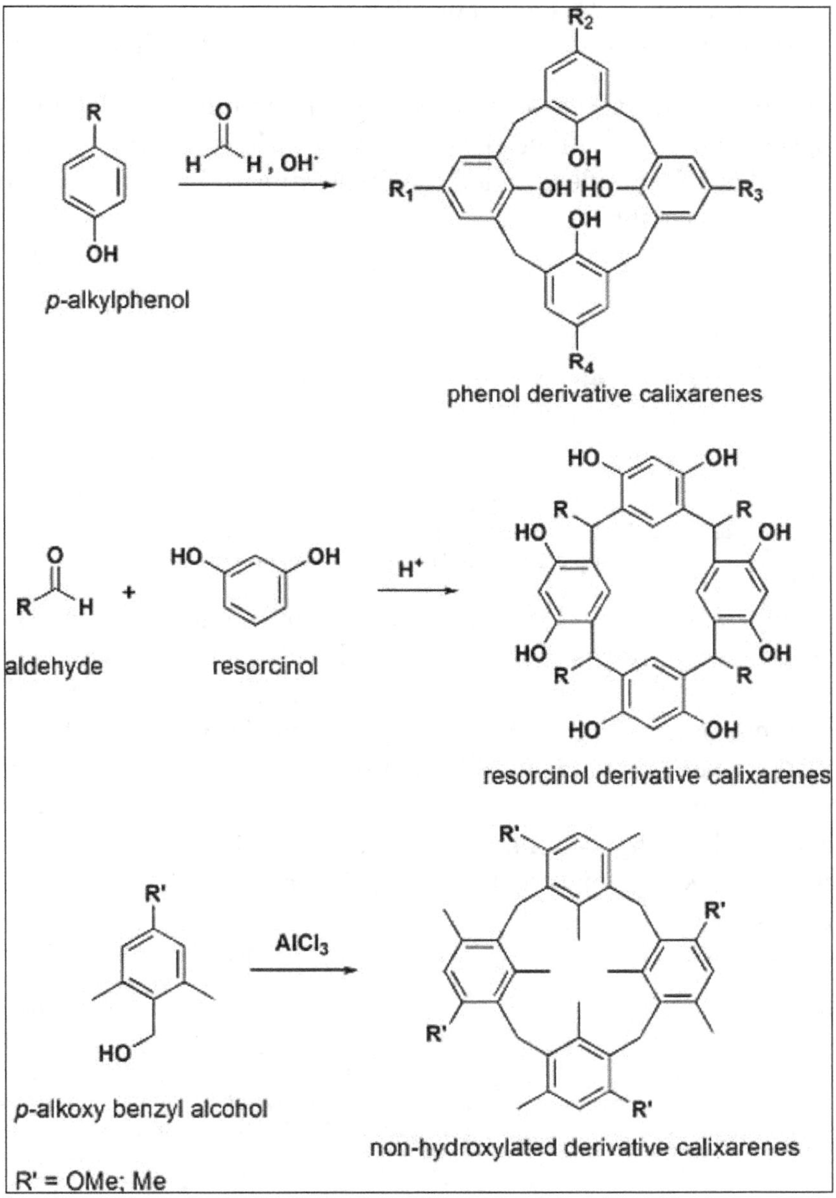

Basic synthesis of calixarene.

Calixarene synthesis can be done through a noncomplicated process. The compound group is commonly synthesized from 4-alkylphenol, especially 4-*t*-buthylphenol, through a one-stage reaction with formaldehyde and NaOH or KOH as the base. The making of calixarenes, known as a base-induced process, results in calixarenes that may vary in their aromatic components. The reaction product between phenols or phenol derivatives and aldehydes is also known as calixarene, while the reaction product between resorcinol or resorcinol derivatives and aldehydes is known as calixresorcinarene. In addition, there are several products that use aromatic compounds in the forms of para-alkoxy-substituted benzyl alcohol derivatives, such as *p*-alkoxybenzyl alcohol, that when reacted with the aid of $AlCl_3$ as a catalyst will result in non-hydroxylated calixarenes. In general, these reactions are illustrated in figure.

The basic structure of calixarenes consists of phenolic units or phenolic derivatives which are repeated and interconnected with Ar-C-Ar bridge to form a hollow cycle. The part with a broader hollow side is called the upper rim, and the narrow hollow part is called the "lower rim" as shown in figure.

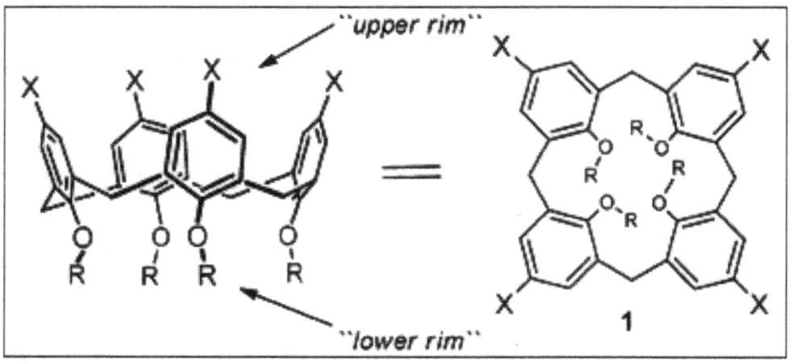

The "upper rim" and "lower rim" of calixarenes.

The various conformations of calixarenes are caused by the free rotation of sigma bonds at methylene, Ar-C-Ar, bridges. For example, calixresorcinarenes have four prochiral centers on carbon atoms at the methylene bridges, causing the compound to form four different diastereoisomers. The relative configurations of the four R substituents at the methylene bridges can form a cis for all R substituents (rccc) or a cis-cis-trans to one R (rcct), cis-trans-trans (rctt), and trans-cis-trans (rtct). The configurations are determined based on their configurations to one of the R substituents on the prochiral center in a clockwise direction. Figure illustrates the relative configurations of the four R substituents.

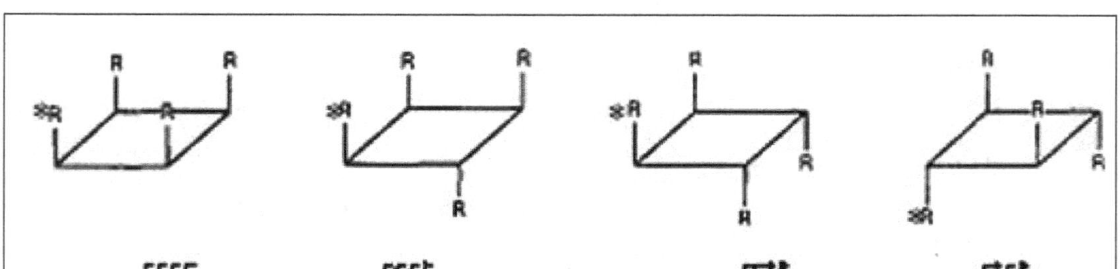

R relative configurations at the methylene bridge.

In reality, the cyclic tetrameric ring of a calixarene is not a planar molecule and can adopt five conformations, namely, "crown" or C_{4v}, "boat" or C_{2v}, "chair" or C_{2h}, "diamond" or

C_s, and "saddle" or S_4. The five conformations of calixarene is shown in Figure. In fact, the rccc isomer where all R groups in the same position, so far only found with the macrocyclic ring adopted the "crown" or C_{4v} and "boat" or C_{2v} conformations. This is caused by the single bond with a tetrahedral shape of R groups so that it can rotate freely. The rccc isomers can form conformation "boat" or C_{2v} with the fourth R possesses an all axial and all-cis configuration, while the macrocyclic ring in a boat conformation. In this case, the fourth R leads into the annulus of the calixarene (endo). It was also reinforced by the result of a condensation reaction between resorcinol and 4-bromobenzaldehyde with an acid catalyst, where produced the rccc isomer of calixresorcinarene in "boat" or C_{2v}conformation. The rctt isomer is only found in "chair" or C_{2h}conformation. In the calculation of energy, the rccc isomer in the crown conformation (C_{4v}) is favored over the rctt isomer in the chair conformation (C_{2h}). Nevertheless, the rccc in boat conformation (C_{2v}) has similar energy to the rctt in chair conformation. The rcct isomer adopts the "diamond" or C_s conformation.

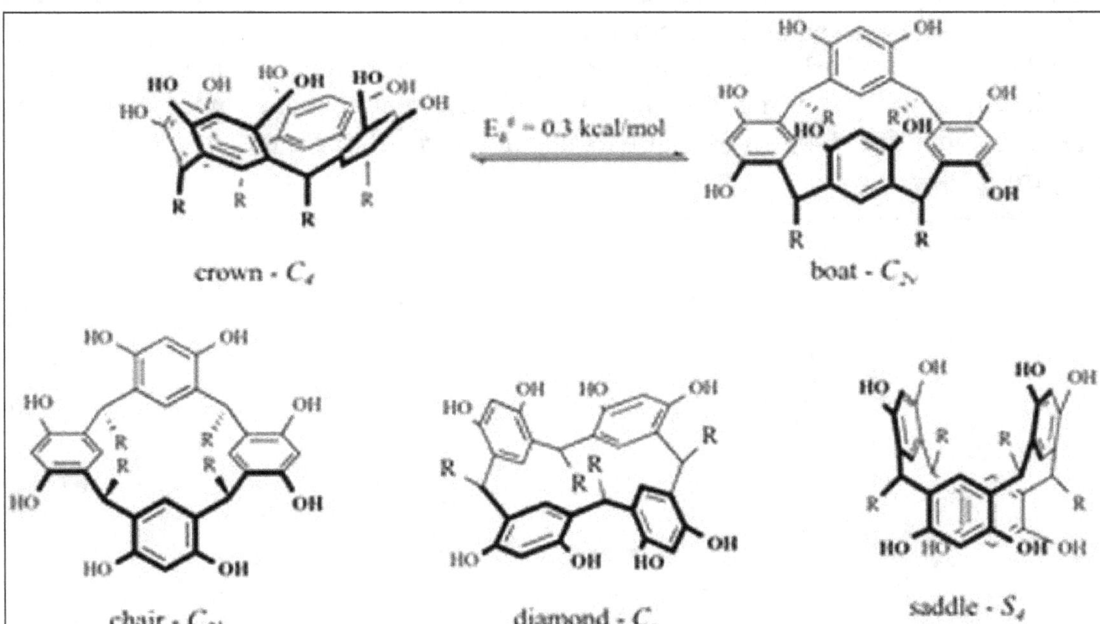

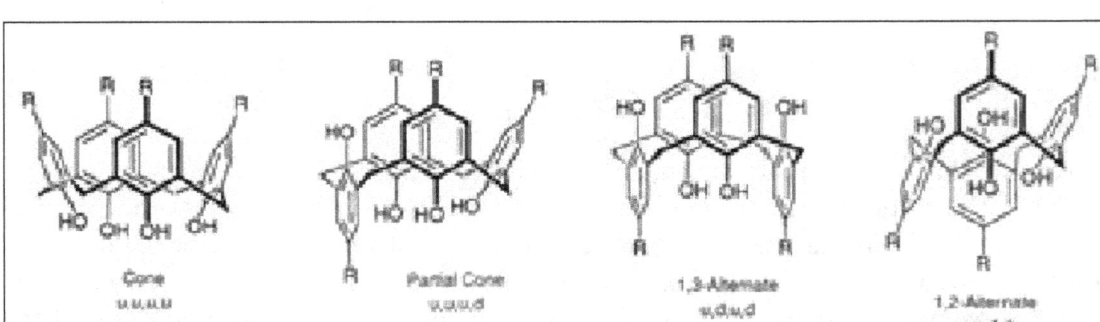

Conformation based on the projection of aryl group in tetramer structure of calixarenes.
Calixarene conformations.

Based on the projection of aryl group on the cyclic tetramer structure of calixarenes, four different conformations can be formed depending on the forms of aryl cluster projection, which can be upward (u) or downward (d), relative to the plane. The four conformations are a cone, partial cone, 1,2-alternate, and 1,3-alternate conformations, each with different thermodynamic stabilities.

In their development, calixarenes have been successfully synthesized not only in the form of the tetramer but also hexamer and octamer. Therefore, the naming of calixarenes is developed by inserting a number inside brackets, showing the number of monomers they possess. For instance, a calixarene containing eight monomers will be written as calixarene. Furthermore, to show the types of phenol used in the synthesis, para substituents are also included in the name. The cyclic hexamer of *p-tert*-butylphenol, for example, is named *p-tert*-butylcalixarene. For resorcinol-derived calixarenes, the name is calix[n]resorcinarene, while the substituents of the methylene carbon are shown by the prefix of "C-substituent." For example, the reaction yield of resorcinol with acetaldehyde is named C-methylcalixresorcinarene, indicating the presence of methyl substituents in the methylene carbon. For a more systematic naming, especially for scientific publication, the basic name of calix[n]arene is maintained, whereas the substituent identity is shown based on its position through the number, such as shown in figure.

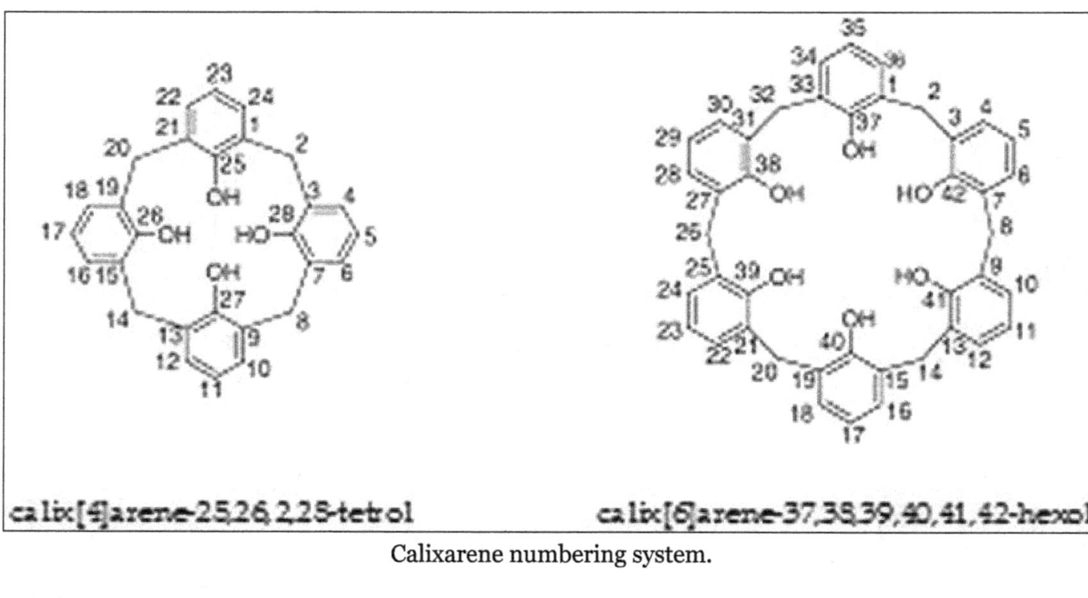

Calixarene numbering system.

Calixarenes have been broadly used, among others, as liquid crystal materials, sensors, catalysts, the stationary phase of chromatography, host molecules, or even semiconductors. As a role in the guest-host system molecule, calixarene and 4-*tert*-butylcalixarene can be the host molecules for compounds such as trifluoromethyl-benzene contained in pesticides. In addition, some calixarene derivatives can also be the host for the dye molecules such as orange I (OI), methylene blue (MB), neutral red (NR), and brilliant green (BG). This finding makes calixarenes as a promising candidate as a host molecule to be used in chemical sensors to measure neutral organic molecules. In addition to being the host molecules for neutral compounds, calixarene and its derivatives were also proven to be a host for the molecular ion compounds such as heavy metal ions. It is no wonder that calix[n]arenes have been greatly synthesized and yielded various oligomers with n = 4–20 according to their purposes. Various modifications of functional clusters have also been developed in order to yield the desired molecular structures, for example, the modification of the aryl group or the aldehyde. The modification can be in the form of the addition of a functional cluster such as alkyl cluster and hydroxyl cluster or the elimination of a functional cluster in the starting materials used or further modification of the calixarene molecules formed. Some of the modifications make the calixarene structure more rigid, thereby forming the expected conformations. To cite an example, the addition of ketone group into aryl group was done to increase the affinity of calixarene compound to the complex formation of alkali cation. The modification can also be made by adjusting the hollow size formed in calix[n]arenes to match the size of a certain molecule both by modifying the functional cluster and adding other components.

Synthesis of Oligomer Calixarenes

Conventional Synthesis Procedures

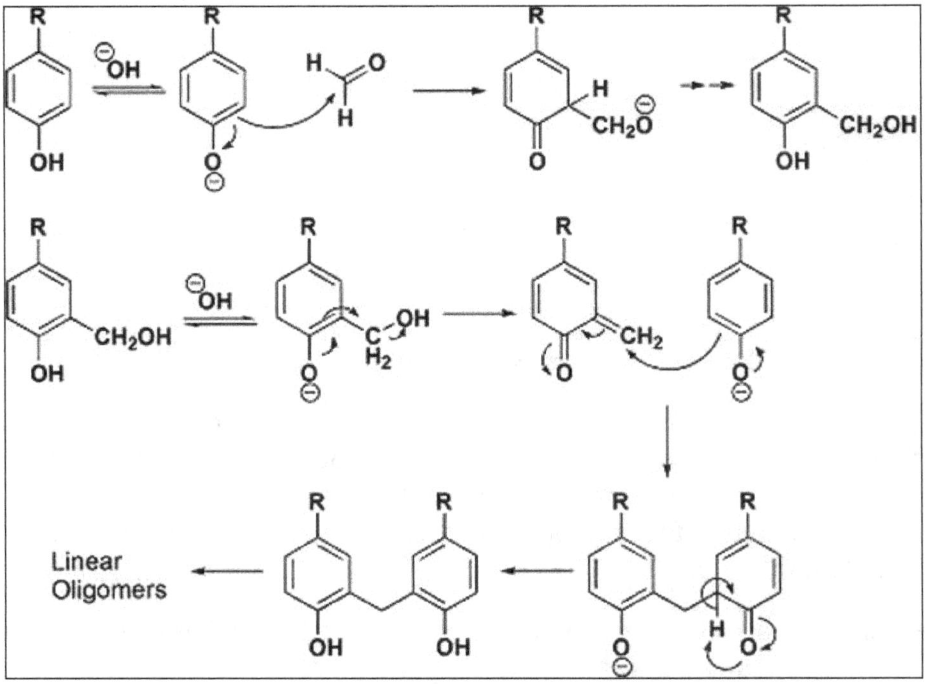

Reaction mechanisms of calix[n]arene formation under basic conditions.

Various calixarene synthesis procedures and their derivations have been largely developed, for example, where the synthesis of calixarene started by mixing *p-tert*-butylphenol, 37% formaldehyde, and NaOH heated at 50–55°C for 45 h. To remove water content, the reaction compound was reheated at 110–120°C for 2 h. In the purification process, the yellow solid yielded was subsequently heated in linseed oil up to 200°C for several hours, and the process yielded tetramers. The modification to this procedure was made by Cornforth by directly heating the reaction compound (*p-tert*-butylphenol, formaldehyde, and NaOH) at 110–120°C for 2 h, thereby yielding a very thick product which was then refluxed with diphenyl ether for 2 h. Once the compound was cold, the product was separated with infiltration and recrystallized to produce white crystals with a melting point of 342–344°C for 50%. The procedure was later on known as the Zinke-Cornforth procedure. The condensation of *p-tert*-butylphenol with formaldehyde under basic conditions (NaOH, KOH) can yield cyclic tetramers, hexamers, and even octamers with a percentage of 85–63%. Figure illustrates the reaction mechanisms of calix[n]arene formation under basic conditions.

The right mechanism to transform linear oligomer into a cyclic oligomer, can be predicted from the formation of hemicalixarene combined with a pair of linear oligomers. Under appropriate conditions, two linear methyl-hydroxylated tetramers combine to form hemicalixarene, followed by the release of water to form calixarene. This process can also be seen in the synthesis of *p*-chlorocalixarene, which is a "2 + 2" condensation product of *p,p*'-dichlorodiphenylmethane, such as illustrated in figure.

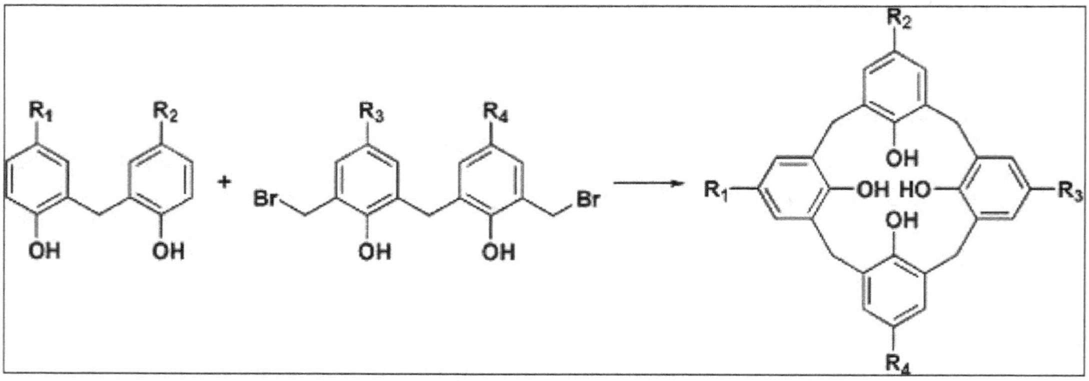

Cyclic tetramer formation through "2 + 2" condensation.

Similarly, calixarene formation is postulated to be the result of a combination of two linear trimers. Based on this postulation, the form of dianion hemicalixarene can be the cause of the large amounts of bases required for hemicalixarene synthesis. The following reaction in Figure, for example, adopts a "3 + 3" condensation that involves two trimers to obtain 9% calixarene.

Gutsche synthesized *p-tert*-butylcalixarene with the effect of basic conditions on the cyclic oligomer produced. The yellow thick paste ("precursor") formed after heating the compound for 1.5–2 h at 110–120°C was diluted with chloroform and washed with aqueous HCl and water several times to neutralize the base. When the precursor containing a base of 0.03–0.04 or equiv. (relative to the number of phenols used) was refluxed for 2 h, it could yield cyclic tetramers. However, the lower the concentration of the base used, the lower the cyclic tetramer formations; on the other hand, a higher concentration of the base will increase the formation of cyclic hexamers. Bases play a role in inducing cyclization and are vital in determining the cyclic oligomer formed. The cations used as the base catalyst can play the role of a "template" which determines the size or number of monomers forming oligomers.

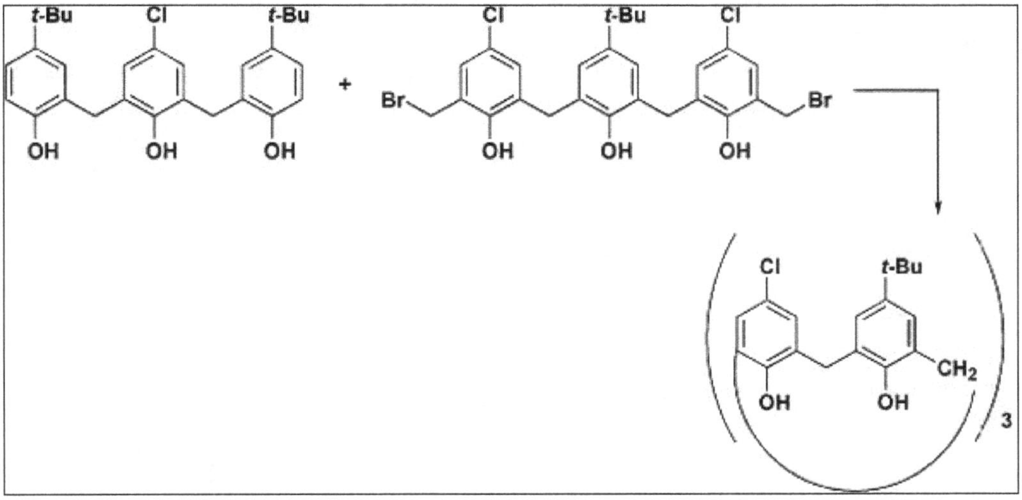

Cyclic hexamer formation through "3 + 3" condensation.

A large cyclic oligomer enables the formation of a larger hollow, and this can be used to absorb heavy metals, as can be found in the synthesis of calixarene involving *p-tert*-butylphenol, paraformaldehyde, and bases (KOH) with a proportion of 0.45 mole relative to the *p-tert*-butylphenol used. The reaction happened for 4 h by refluxing it in *p*-xylene. The compound was separated with liquid-liquid extraction using chloroform. The organic phase obtained was then evaporated and recrystallized with methanol to yield 65.47% white solids with a melting point of 370–372°C. The analysis of the product showed that it is a cyclic hexamer of calixarene.

Condensation of resorcinol with formaldehyde under acidic conditions formed a linier or cross-linked polymer between formaldehyde with resorcinol. This reaction also did not produce calixresorcinarene under alkaline conditions. It is not surprising since resorcinol and formaldehyde are highly reactive so that the reaction becomes uncontrollable and does not produce a cyclic oligomer. Base conditions can be applied to the synthesis of calixresorcinarene from 2-substituted resorcinols and formaldehyde. Nevertheless, resorcinol used should be substituted electron-withdrawing group at position 2 as shown in Figure. Electron-withdrawing group at position 2 will reduce the reactivity of resorcinol so that the reaction will be controlled and it will be suitably carried out under alkaline conditions. The presence of electron-withdrawing substituents such as nitro group (–NO_2) facilitates condensation reactions that occur by directing the reaction site at the meta position of the nitro group. The reaction rate decreased from electron-withdrawing substituent nitro > acetyl > carboxylate > H due to the ability of the groups to deactivate the resorcinol. In contrast, the reaction of resorcinol substituted electron-donating group such as 2-methylresorcinol or phloroglucinol (=benzene-1,3,5-triol) with formaldehyde under base conditions did not produce a derivative of calixresorcinarene. Therefore, the acidic conditions are more suitable when used resorcinol substituted electron-donating group in position 2. The condensation reaction between formaldehyde and 2-nitroresorcinol and its analogs seems to take place in the irreversible reaction, so it is unknown whether to form a tetramer, pentamer, or hexamer.

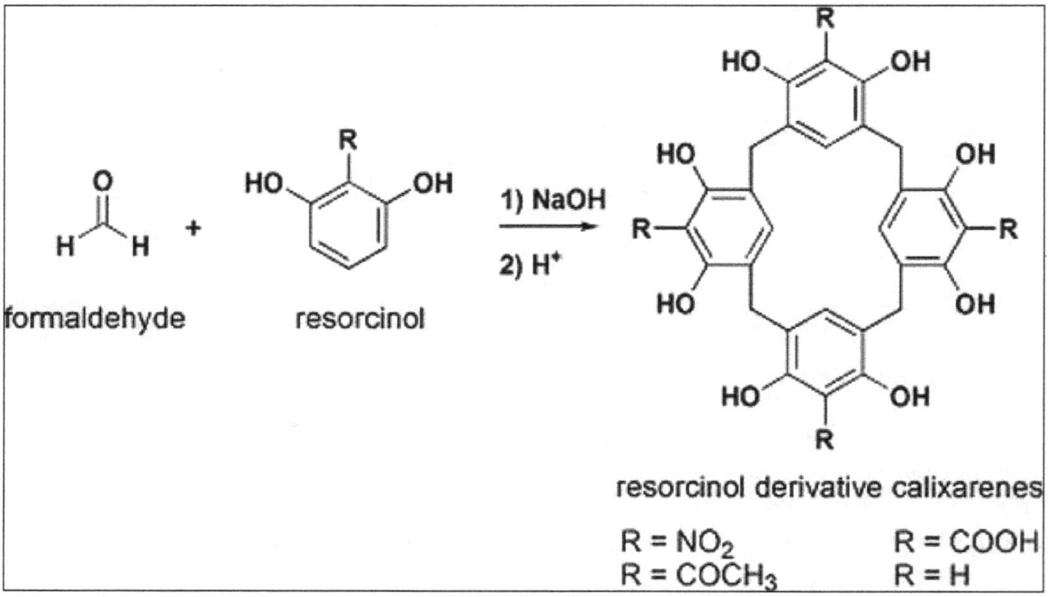

Synthesis of resorcinarenes from 2-substituted resorcinols and formaldehyde in basic conditions.

Besides adding bases, the synthesis of calix[n]arenes can also be done under acidic conditions. One such example is found in the procedure of calix[n]resorcinarene synthesis developed by Niederl-Hogberg, in which resorcinol was used as an aromatic unit in the reaction. Resorcinol is more reactive compared to phenol. Resorcinol was reacted with acetaldehyde by adding hydrochloric acid. The reaction took place at 80°C for 16 h and yielded C-methylcalixresorcinarenes for 75%. The reaction follows the following mechanism which is illustrated in Figure.

In the condensation reaction catalyzed by acids, kinetic and thermodynamic factors compete with each other. The kinetic factor controls the oligomer rate produced, while the thermodynamic factor controls the cyclic structure formation. In this reaction, controlling the thermodynamic factor will be very determining for the formation of a small cyclic oligomer, such as the formation of calixresorcinarene. To control the thermodynamic factor, less reactive aldehydes can be used to slow down the reaction rate. If the aldehyde used is very reactive (such as formaldehyde), the reaction will happen very quickly so that the kinetic factor will dominate and the majority of the products formed will be polymers with linear chains. A modified aldehyde was used in the synthesis reaction of methoxyphenylcalixresorcinarene from resorcinol with *p*-anisaldehyde with the addition of HCl. The reaction happened employing reflux for 30 h, yielding 4-methoxyphenylcalixresorcinarene as much as 91.54%.

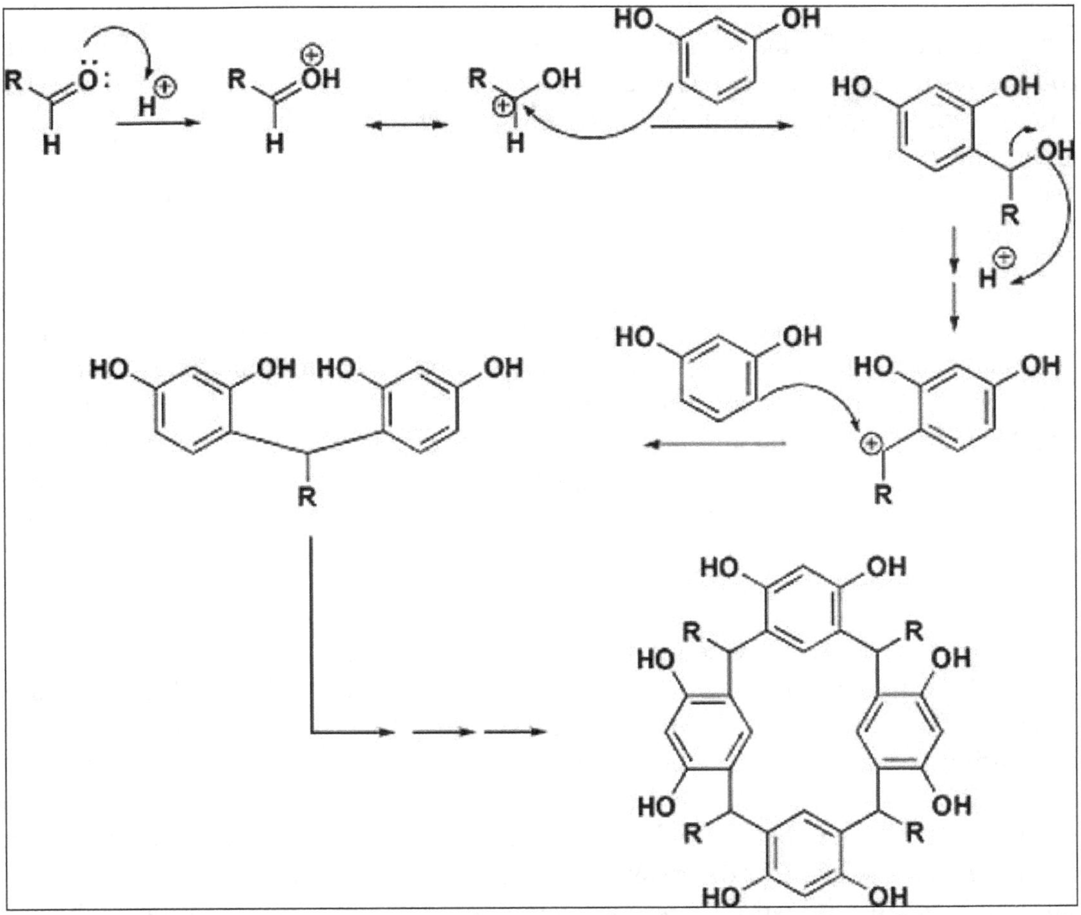

Reaction mechanism of calix[n]resorcinarene formation under acidic conditions.

The synthesis of calixresorcinarene from resorcinol with acetaldehyde was done by Petrova et al. in 2012. The reaction was made by varying the amounts of hydrochloric acid used. The reaction occurred at 75°C for 5–24 h. The reaction product was dependent upon the amounts of hydrochloric acid used. Increasing the amount of acid can cause the formation of linear oligomers. The amount of acid catalyst effective in this reaction was 4% of the resorcinol. In addition, the polarity of the solvents used affected the composition of the yielded product. An increase in the polarity of the solvent caused an increase in the oligo cyclization. Calixresorcinarene with rccc conformation was quantitatively formed very well (76.89%) when 30% water-ethanol solvent was used.

Calixresorcinarenes synthesized through a condensation reaction between resorcinol and an aliphatic or aromatic aldehyde under acidic conditions yield products in the form of cyclic tetramer calixresorcinarenes that theoretically can contain four diastereoisomer products. However, in reality, only three have been identified to be formed by the reaction with an aliphatic aldehyde and two from the reaction of an aromatic aldehyde with resorcinol. The products with crown conformation with all side chains in the axial position (rccc) are more desirable and stable thermodynamically. Hogberg reacted resorcinol with acetaldehyde by adding hydrochloric acid. The reaction occurred at 75°C for 1 h and yielded sediment that was subsequently acetylated. The acetylated product was in the form of two isomeric cyclic tetramers, one of which formed cis (rccc) conformation of up to 47%, while the other product formed rctt conformation with a percentage yield of 13%. This indicated that products with all side chains in the axial position (rccc) are more desirable and stable thermodynamically. selectively increasing the alcohol proportion in the solvent mixtures used and increasing the reaction temperatures and time.

Several synthesis methods of calix[n]arenes and their derivatives have been largely developed to produce high yields. However, the methods require heating at high temperatures and a long reaction time. This requirement is disadvantageous, ultimately when the synthesis is done on a large scale. The required solvents will be in a very large amount, and the reaction time needed will be much longer. This makes the synthesis of calix[n]arenes and their derivatives not environmentally friendly.

Green Synthesis Procedures

Various approaches to calixarenes and their derivatives have been attempted, one of which is by implementing the concept of "green chemistry." The synthesis of calixarenes and their derivatives requires heating at high temperatures for quite a long time. The heating is usually done using conventional heating, where heat transfer occurs conductively. The method is relatively slow and inefficient to transfer energy into the reaction system, as it depends on the thermal conductivity of the materials used; consequently, the temperature of the reaction vessel will be greater than that in the reaction compound. Irradiation with a microwave can produce more efficient internal heating because microwave energy penetrates into the containers so that it can directly hit the reaction compound, and energy transfer becomes more efficient. The phenomenon will depend on the ability of the molecules (solvents or reagents) to absorb microwave energy and change it into heat. The thermal energy is obtained when the polar molecules try to align themselves with an electric field resulted from microwave irradiation. The amount of heat produced depends on the dielectric properties of the molecules. Dielectric constants indicate the ability of the molecules to get polarized by the electric field. The electric field received by the molecules causes reorientation of the dipolar molecules which in turn increases polarity and reactivity by reducing the activation energy. Consequently, the reaction can happen more quickly.

Organic synthesis using microwave technology has been largely employed, as it can accelerate the reaction time as a consequence of an increase in heating that cannot be achieved with conventional heating. This is what happened in Baozhi's synthesis of *p*-methylcalixarene. A reaction compound consisting of formaldehyde, *p*-methylphenol, and KOH was heated with microwave for 2–8 min (output power at 100%).

The required reaction time at this stage was faster than that when using conventional heating, which could take up to 2 h. After solids were formed, reflux was done for 40 min using dimethylbenzene. When cooled down, white crystals with a melting point of 361–362°C were yielded for up to 81.7%. The synthesis of *p-tert*-butylcalixarene was done with the same procedure using the microwave as well, in which the compound was radiated for 4–8 min. To yield a larger cyclic reaction product, the amounts of bases used were increased with a longer reflux time. Bases and reflux time play a role in cyclic formation. The same phenomenon was observed when calixarene synthesis was done using resorcinol, which is known to be more reactive than phenol. The reaction product in the form of cyclic tetramer calix[n]resorcinarene was obtained after the compound was heated with conventional heating for 5–10 h, with a reaction yield of around 60–90%. However, the reaction could be faster when the heating method was changed with microwave. One reaction only required 3–5 min with a greater reaction percentage yield, namely, around 80–90%. This finding was put to use in the synthesis of thiacalix[n]arene with n = 4–8.

Aldehyde modification can also be done by using the abundant natural products such as vanillin that can be derived from vanillin plant (*Vanilla planifolia*), cinnamaldehyde from cinnamon oil, and anisaldehyde from Indonesian fennel oil. The synthesis procedures using these materials have been developed for the synthesis of calix[n]arenes and their derivatives with "green chemistry" approach. The three compounds were reacted with resorcinol to yield cyclic tetramers, calixresorcinarenes. The synthesis reaction of calixresorcinarenes can occur for 15–30 h using conventional heating. Sardjono et al. developed a microwave-assisted synthesis procedure for calixresorcinarenes, where resorcinol was reacted to vanillin or cinnamaldehyde. The calixresorcinarenes derivatives yielded could be up to 97.9% in 8 min with a ratio of resorcinol to vanillin or cinnamaldehyde was 1:1, and the microwave power used 332 W. Another calixresorcinarenes derivatives, namely C-anisalcalixresorcinarene could also be yielded for up to 99.5% in 5 min from the reaction of resorcinol with anisaldehyde at the ratio of 1:1.2 and 264 W microwave power.

Vanillin in the synthesis of calixresorcinarene derivatives was also employed by Nakajima and Kobayashi, although they still used the conventional heating method (without a microwave). The reaction compound comprised resorcinol and aldehyde (vanillin, syringaldehyde, or *p*-hydroxybenzaldehyde) with a mole ratio of 1:1 and hydrochloric acid heated at 70°C and stirred for 12 h. The reaction yield was then washed with hot water and purified using methanol and ethanol. The third reaction yield formed cyclic tetramers, namely, vanillin calixarene for 32%, syringaldehyde calixarene for 28%, and *p*-hydroxybenzaldehyde calixarene for 37%.

Calixarene synthesis commonly needs a lot of solvents which results in a large amount of waste. Therefore, a synthesis approach by minimizing the use of solvents has been attempted and proved to be more effective. The solvent-free reactions that have been tried, in addition to requiring a shorter amount of time, used less energy and could produce better and more effective and selective reactions than the conventional method. Solvent-free reactions have also been used for calixresorcinarene synthesis, which is a condensation reaction between aldehyde and resorcinol. The method has also been widely used in several organic reactions and sometimes gives better results than the reaction method in solvents.

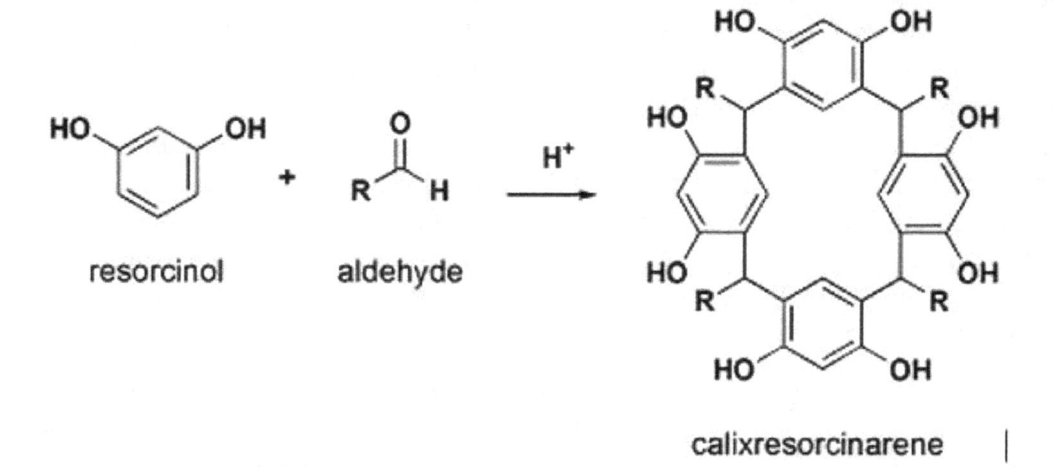

Aldehyde	R	Calixresorcinarene
Vanillin (4-hydroxy-3-methoxyphenylbenzaldehyde)	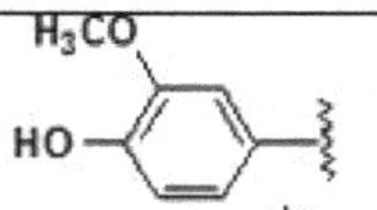	C-4-hydroxy-3-methoxphenil calix[4]resorcinarene
Cinnamaldehyde (4-methoxyphenylbenzaldehyde)	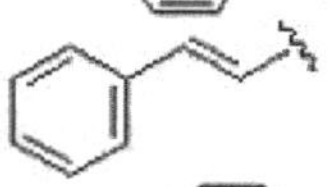	C-2-phenylethenil calix[4]resorcinarene
Anisaldehyde	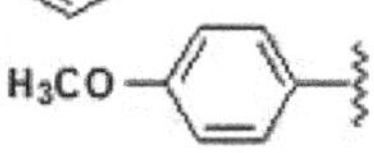	C-4-methoxyphenyl calix[4]resorcinarene

Synthesis reaction of calixresorcinarene from resorcinol with various aldehydes.

In calixresorcinarene synthesis, the use of solvent-free method was carried out by mixing aldehyde and resorcinol (1:1) with a certain amount of acid catalysts and then ground simultaneously. Although all reagents were in solid forms, when mixed they could form a thick liquid or paste reaction compound. Isolation of the reaction product can be carried out by washing with water to remove the acid catalyst and recrystallize with the right solvent. The method was proven to effectively yield calixresorcinarene with various aldehydes. In this method, no external heating is needed; only consistent scouring is needed. The scouring can be done in order to reduce particle sizes so that the compounding will be efficient. Because the reaction forms a thick liquid, not powder compound, then scouring at high intensity is not necessary.

Using the solvent-free method, the reaction occurs in mere minutes, much faster than the reaction time using the conventional method, which requires hours or even days. The percentage yield of the solvent-free method is often greater than that of the conventional synthesis method. The synthesis of calixresorcinarenes commonly yields isomeric products, with two forms of isomer, namely, cis-cis-cis (rccc) and cis-trans-trans (rctt) with "crown" (C_{4v}), "chair" (C_{2h}), or "boat" (C_{2v}) conformation. Theoretically, without the effect of solvents, rccc isomers with "crown" (C_{4v}) conformation are more desirable than the rctt isomers with "chair" (C_{2h}) conformation. Besides reduced reaction time, synthesis with this method uses a small amount of catalyst and does not produce solvent waste, hence meeting the principles of "green synthesis".

The condensation reaction between aldehyde and resorcinol (1:1) with *p*-toluenesulfonate acidic catalyst (5%) occurs for several seconds up to several minutes and yields derivatives of calixresorcinarenes with rccc and rctt isomers. Four out of the six reactions were reported by Roberts et al. to yield more rctt isomers with "chair" (C_{2h}) conformation than rccc isomers with "crown" (C_{4v}) conformation, with a ratio of 2:1. However, in one of the reactions, more rccc "crown" (C_{4v}) isomers were yielded for up to 95%. The distribution of isomer products is very possibly related to solubility that also plays a role in the reaction with the solvent-free method. In this reaction, increased reaction or heating time of the initial materials up to 85°C for 5 h did not cause an increase in the amount of rccc isomers yielded with "crown" (C_{4v}) conformation.

Employing the same method, Firdaus et al. reacted vanillin with resorcinol (1:1) and yielded C-4-hydroxy-3-methoxyphenylcalixresorcinarene for 52%, with a ratio of rccc isomers to rctt isomers of 1:5. However, when the aldehyde used was replaced with *p*-anisaldehyde, 63% of C-4-methoxyphenylcalixresorcinarene was yielded with rccc isomers two times greater in number than rctt isomers. The distribution of the isomer products showed that each aldehyde has such a different characteristic that it can yield a different product composition. This difference may also be related to the solubility of the aldehyde used in the reaction compound. In addition, the reaction product demonstrated that rccc isomers with "crown" (C_{4v}) conformation are more desirable.

The solvent-free method is significantly faster than the conventional synthesis method. This can also be seen in the synthesis of calixpyrogallolarene that is commonly done with condensation reaction of pyrogallol and aldehyde using alcohol solvent and acid catalyst and yields in rccc "cone" isomers. In the synthesis of calixpyrogallolarene using the solvent-free method, an amount of aldehyde was added drop by drop to pyrogallol (1:1) with *p*-toluenesulfonate (3.7%) catalyst and scoured constantly. The reaction yielded vulnerable white solids in 2 min. The solids were easily scoured into soft yellow powder in 5 min. As was the case with the synthesis reaction of calixresorcinarene derivatives, no heating was used in this synthesis.

The synthesis of C-(*p*-substituted phenyl) calixresorcinarenes was carried out to see the interaction between host and guest using DMSO by mixing resorcinol with halogen-substituted benzaldehyde in para positions (1:1) under acidic conditions with an addition of 5% solid *p*-toluenesulfonate. The synthesis yielded a product at a percentage of 90–95% by scouring without using any solvents.

In the synthesis of calixresorcinarenes from resorcinol and aromatic aldehyde, Lewis acid is frequently used, such as BF_3-OEt_2, $AlCl_3$, and $SnCl_4$. For example, in the synthesis of resorcinarene *O*-acetates from 1,3-di(alkoxycarbonylmethoxy)benzenes with aromatic or aliphatic aldehydes (1:1) can be efficiently done at room temperature with the addition of BF_3-OEt_2 (boron trifluoride etherate) as a catalyst in order to produce high yields. However, the amount of acid needed in this reaction is quite large (50–200% mole). Certainly, this is not good for the environment and safety, considering the sensitivity of Lewis acid to water that can yield side products in the forms of acidic oxides and metal oxides. Furthermore, the process is not efficient in terms of atom economy. The reaction also can yield two diastereoisomers. Lanthanide (III) triflates are selective acid catalysts that have been widely used for the formation of carbon-carbon and carbon-heteroatom bonds. Different from the common Lewis acid that is frequently used in stoichiometric numbers, lanthanide (III) triflates can be used in various reactions in catalytic numbers in the presence of solvents such as THF, DMSO, DMF, MeCN, and water. In addition, the stability and solubility of this catalyst in water make the compound easily separated and reused. One of the examples of lanthanide (III) triflate compounds that can be used as a catalyst is Ytterbium (III) triflate nonahydrate $\{[Yb(H_2O)_9](OTF)_3\}$. In the condensation reaction of resorcinol with various aldehydes (1:1), to produce derivatives of calix[n]resorcinarenes only needs 8% mole of Ytterbium (III) triflate nonahydrate as a catalyst. Although the amount of the catalyst is small, the reaction product yielded can reach up to 71–94% and this reaction is stereoselective. The reaction is done by refluxing for 48 h with ethanol as solvent.

Tungstate sulfuric acid (TSA) can also be used in the synthesis of calixresorcinarene derivatives. In the reaction of an aldehyde with resorcinol at 120°C, the addition of TSA could produce around 78–95% in 25–80 min. The use of TSA as an acid catalyst is also more environmentally friendly because this acid can be recycled and reused for up to four reaction cycles with a decrease in reaction product yield around 3–11%.

Green Chemistry in Organic Synthesis

A green chemistry campaign was initiated so that organic substances could be produced in an environmentally clean manner. The basic principles guiding this new synthetic strategy are simply summarized in the following goals:

- Reducing potential waste production by utilizing atom economical processes, which minimize the use of solvents.
- Reducing the use of hazardous reagents or the production of hazardous by-products.
- Utilizing energy economically and looking at energy sources that do not rely on fossil fuel combustion.

- Minimizing time so that energy consumption is reduced.
- Maximizing yields by utilizing catalysts when appropriate.

Synthetic Approaches

Solvent-free Reactions

Long before the recognition and development of green chemistry concepts, we utilized methodologies employing solvent-free reactions for the production of 1,2,4-triazole carboxamides, recently patented as cannabinoid receptor blockers.

Mixtures of 1 and a variety of amines when stirred in refluxing water yield either 2 or 3 depend on the nature of the amine utilized. Recently we have also observed that 4 undergoes rearrangement to form 5 under these reaction conditions.

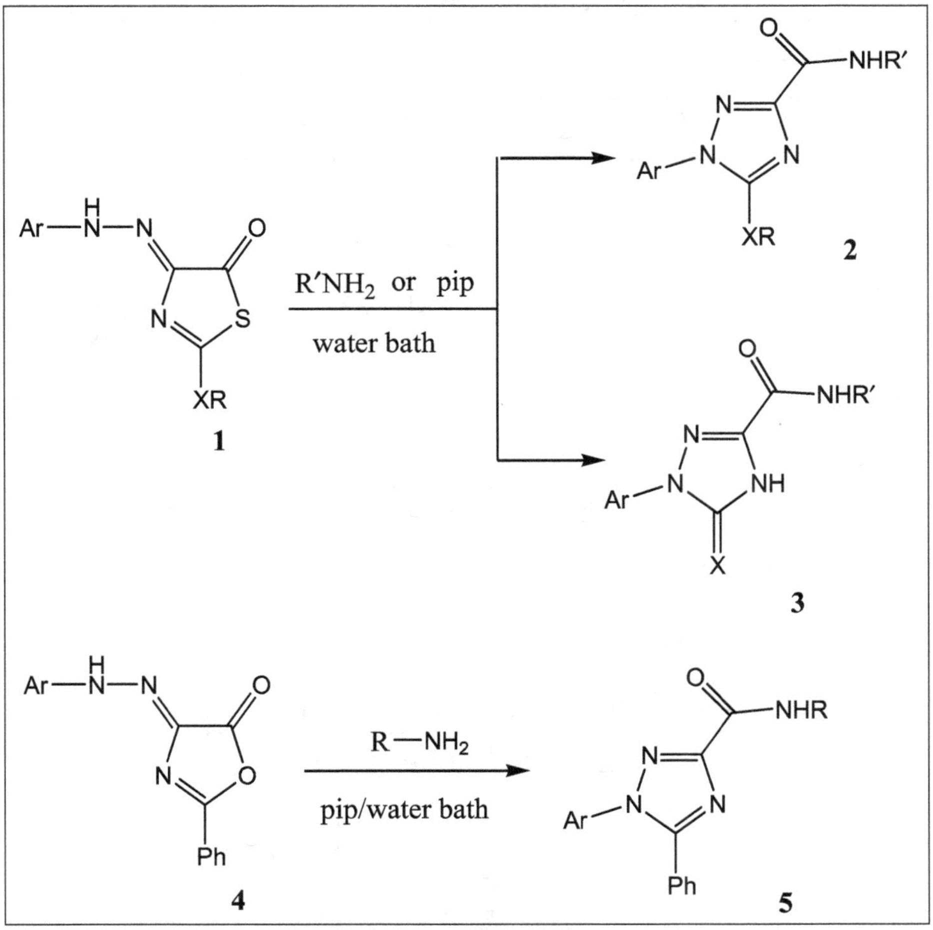

Rearrangements of compounds 1 and 4.

We have also described the conversion of 6 to 7 via reaction with ethyl cyanoacetate in the presence of a catalytic amount of both AcOH and ammonium acetate.

Recently, a green, highly efficient, three-component process for the synthesis of thiazolo[3,2-a]pyridine derivatives 8 and 9 was described. Accordingly, reactions were observed to occur between malononitrile, aromatic aldehydes, and 2-mercaptoacetic acid in respective molar ratios of 2:2:1.2 and 2:2:2.1 and in the presence of catalytic amounts of piperidine in the absence of solvent to produce 8 and 9. The structures of the products of these reactions were established using HMBC and heteronuclear multi-quantum correlation (HMQC) techniques. The findings contrast to the reported formation of pyronothiazoles 10 in similar reactions.

Table: Yields of Compounds 2a=c and 3a-c.

Compounds	X	R	R?	Ar	Yield%
2a	S	$C_6H_5CH_2$	C_6H_5	C_6H_5	80
2b	S	$C_6H_5CH_2$	C_6H_5	*o*-$CH_3C_6H_4$	65
2c	S	$C_6H_5CH_2$	C_6H_5NH	C_6H_5	73
3a	S	–	CH_3	C_6H_5	80
3b	S	–	CH_3	*o*-$CH_3C_6H_4$	78
3c	S	–	Piperidino	C_6H_5	85

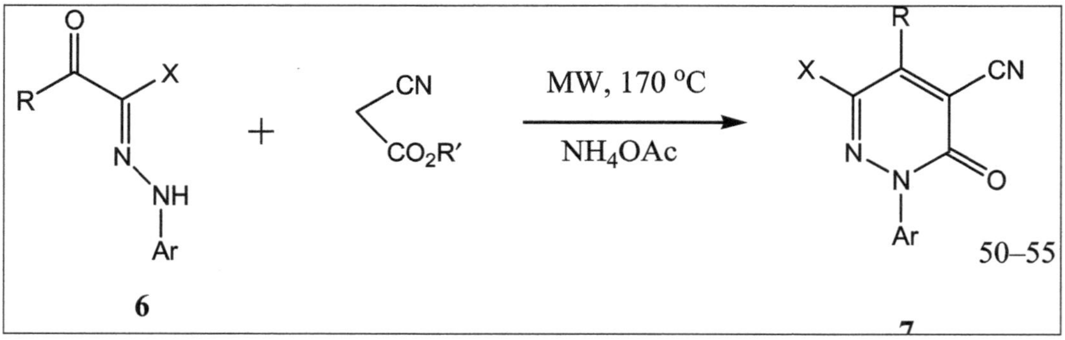

Syntheses of pyridazines.

Table: Yields of Compounds 8 and 9a-c.

Compounds	Ar	Yield%
8a	C_6H_5	85
8b	*p*-BrC_6H_4	82
8c	*p*-$CH_3OC_6H_4$	88
8d	*m*-$O_2NC_6H_4$	82
9a	C_6H_5	90
9b	*p*-$CH_3OC_6H_4$	92
9c	*m*-$O_2NC_6H_4$	87

Enaminones 11 are produced via reactions of methyl ketones with dimethylformamide dimethylacetal (DMFDMA) in refluxing toluene, but the yields of these processes are rather poor. Heating methyl ketones with DMFDMA in absence of solvent for 48 h, however, leads to the formation of 11 in much improved yields. A similar methodology has been developed for the synthesis of 15 from 10, triethylorthoformate and morpholine or piperidine 14a,

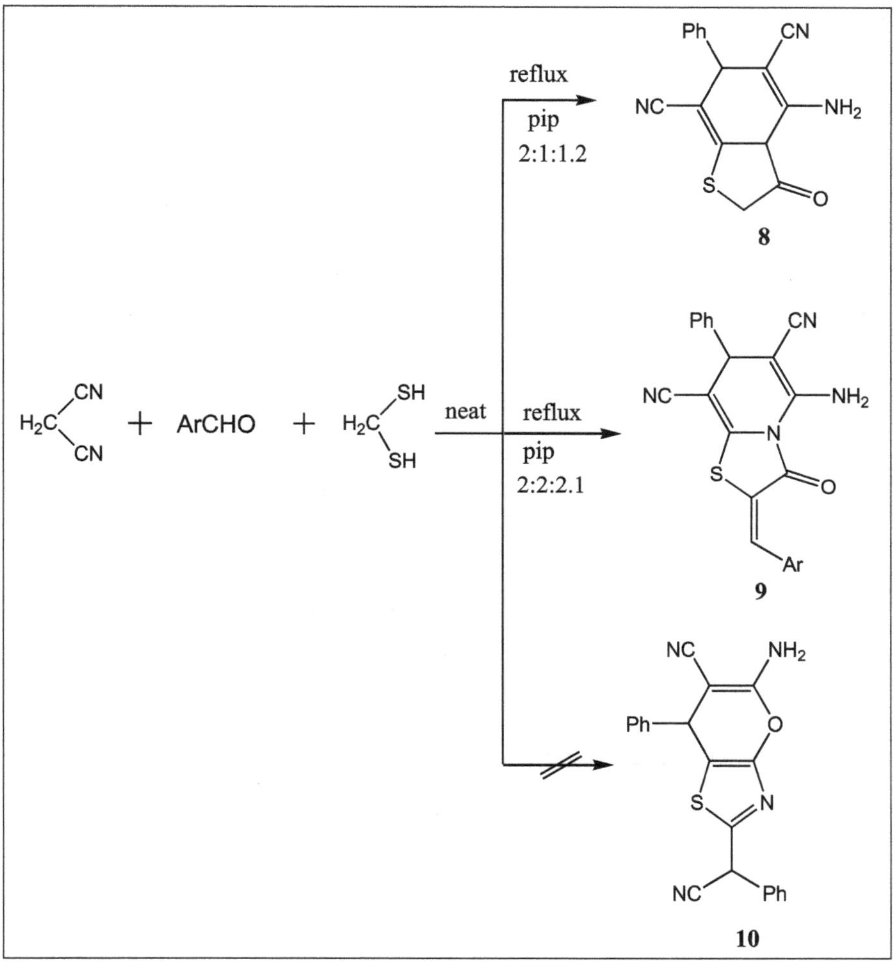

Syntheses of compounds 8a-d and 9.

Utilization of Microwaves and Ultrasound as Energy Sources

Since Gidey and Gidey described the utility of microwave irradiation produced in domestic ovens as an energy source for conducting organic reactions in short time periods, the utility of this technique in organic synthesis has gained great attention. This is especially true when microwave ovens that are specially designed for chemical reactions were constructed.

Subjecting equimolecular mixtures of 2-aminopyrimidine 22, aromatic aldehydes 23, and active nitriles 24 in the presence of piperidine as a catalyst to microwave irradiation leads to the formation of either 4-amino isomer 25 or 2- amino isomer 26, depending on the active methylene compound employed.

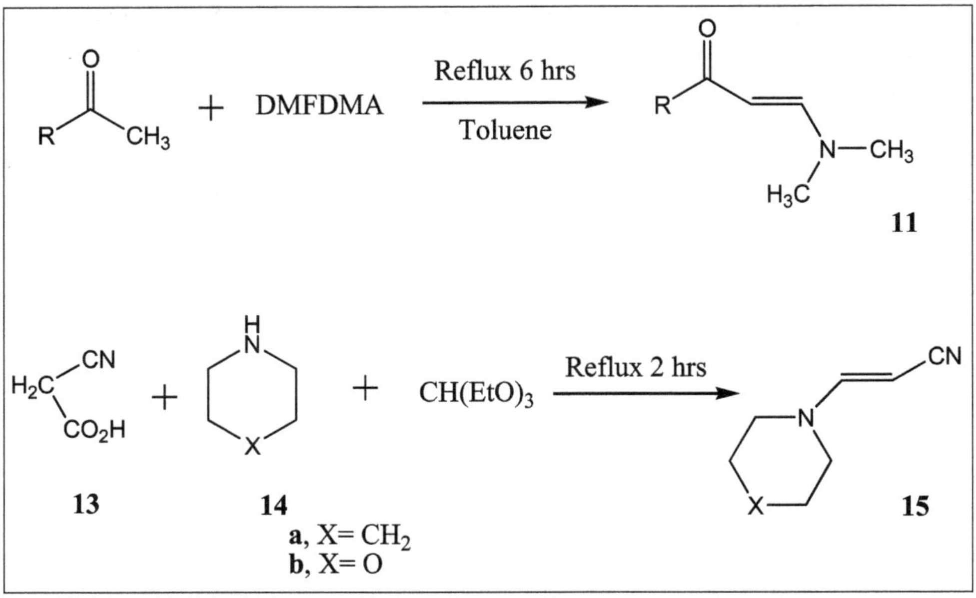

Syntheses of compounds 11 and 15.

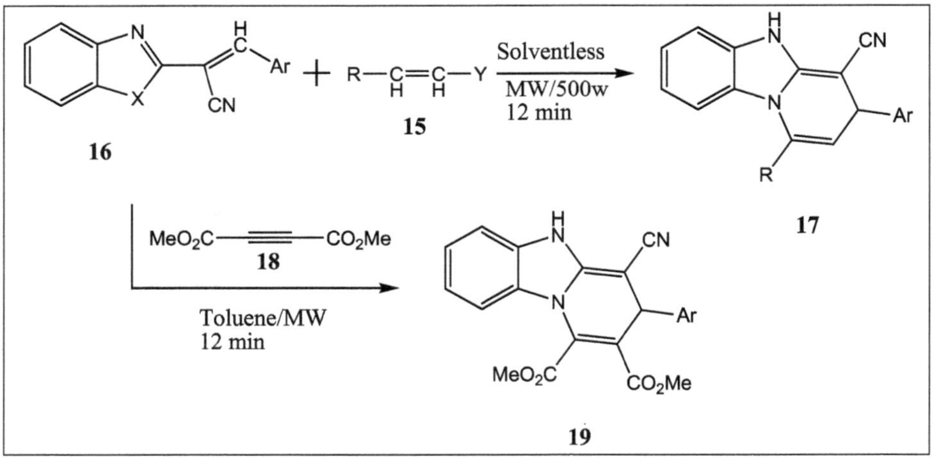

Syntheses of compounds 17a–d and 19a–b.

Table: Yields of Compounds 17a–d and 19a–b.

Compounds	R	Ar	X	Yield%
17a	H	C_6H_5	$CONH_2$	80
17b	H	p-$CH_3OC_6H_4$	$CONH_2$	82
17c	C_6H_5	C_6H_5	NO_2	88
17d	C_6H_5	p-$CH_3OC_6H_4$	NO_2	90
19a	–	C_6H_5	–	85
19b	–	p-$CH_3OC_6H_4$	–	88

Benzoylacetonitrile 31 has long been known to yield the self-tricondensation product 32 upon refluxing in pyridine. Abdelrazek and Michael have subsequently reported that

33 is actually the reaction product. pyridine leads to the formation of the aniline derivative 35 that undergoes hydrolysis to afford 36. 32 is the product formed. Moreover, instead of requiring the reaction to be carried out in refluxing pyridine for at least 5 h, we found that heating neat 19 in a microwave oven for 5 min afforded 32.

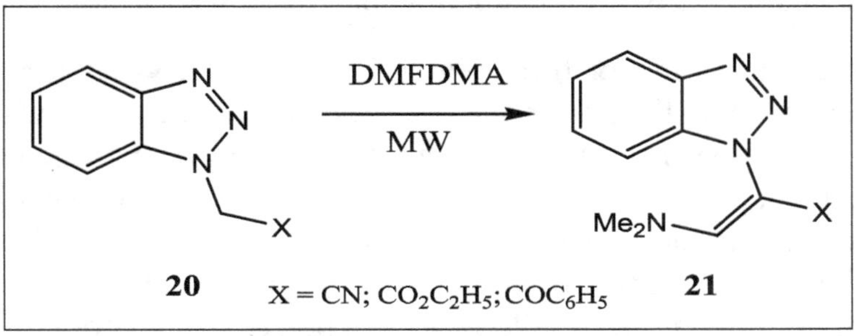

Synthesis of compound 21.

Recently, a novel, simple, and efficient method for the synthesis of polysubstituted diaminobenzonitriles has been developed in our group. The process involves the reaction of 1,1,3-tricyano-2-aminopropionitrile with nitro olefins under controlled microwave irradiation conditions.

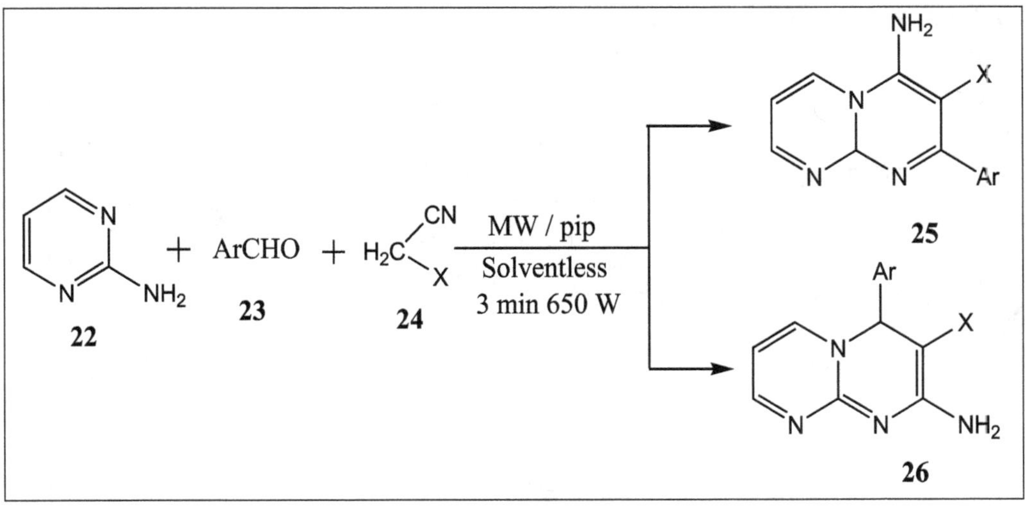

Syntheses of compounds 25a-d and 26ab.

Table: Yields of compounds 25a-d and 26a-b.

Compounds	X	Ar	Yield%
25a	CN	C_6H_5	85
25b	CN	p-ClC_6H_5	83
25c	$CO_2C_2H_5$	2-Furyle	81
25d	$CO_2C_2H_5$	m-$NO_2C_6H_4$	84
26a	C_5NH_2	p-$CH_3OC_6H_5$	82
26b	C_5NH_2	m-$NO_2C_6H_4$	83

Thus, when equimolar amounts of 37 and (E)-2-nitrophenylarenes 38 in dioxane are irradiated using a microwave lab station at 100 8C in the presence of piperidine (2 drops), the corresponding polysubstituted diaminobenzonitriles 39 are generated rather than the corresponding 2-cyanomethylpyridine derivatives 40. The structures of 39 were established using 1 H NMR and C NMR spectroscopic analysis.

Microwave promoted heating mixtures of 2-arylhydrazonals 41 with acrylonitrile 42a or methyl vinylketone 42b in the presence of 1,4-diazabicyclo[2,2,2]octane (DABCO) for 5 min in the absence of solvent results in the formation of 45 via intermediates 43 and 44 that can be isolated in some cases. Reactions 41 (XCOR) with hydroxylamine hydrochloride in acetic acid in presence of sodium acetate under microwave irradiation conditions afforded 46, which is an ecofriendly route to this substance that serves as an alternative to Shawali et al.'s synthesis, which utilizes the reaction of 47 with cyanide ion, or the use of Elnagdi's coupling method of aroylacetonitrile with aryldiazonium salts Notably, this reaction can be conducted on a 0.01 mol scale giving a reported yield of 93% for 45.

Recently Al-Zaydi et al. described a procedure for the formation of pyridines 48 that involves microwave or ultrasound irradiation of a mixture of 49 and an amine followed by treatment of the intermediate 50 with ethyl cyanoacetate also under either microwave or ultrasound irradiation.

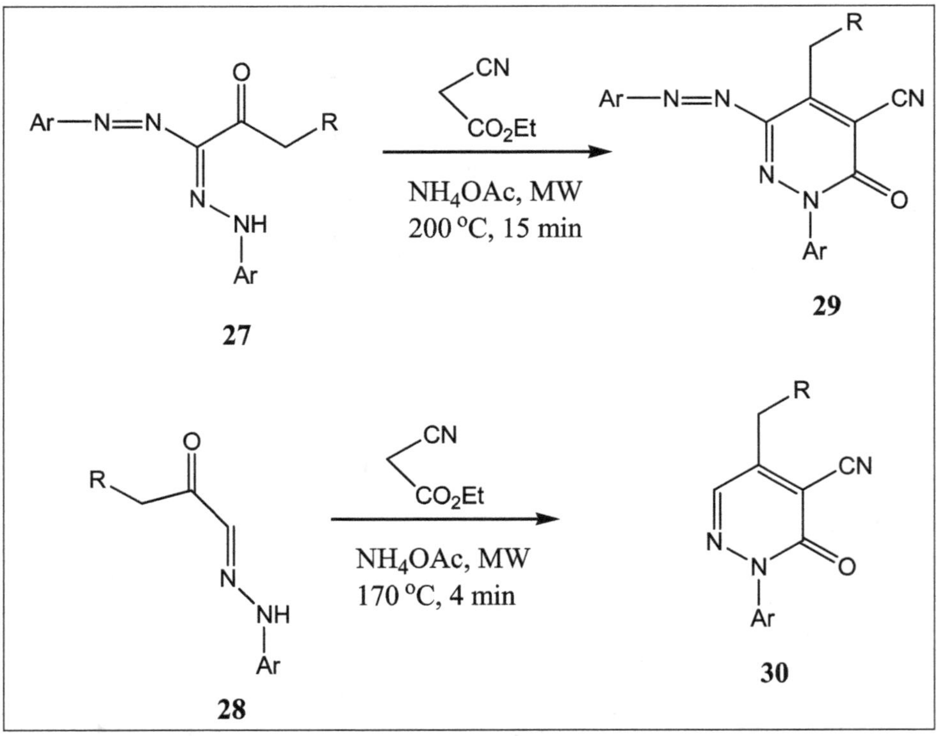

Syntheses of compounds 29ab and 30.

Generally, microwave irradiation proved to be more efficient in producing the desired pyridines in higher yields. Pyridones 48 couple with aromatic diazonium salts to yield arylhydrazones 51, which react with elemental sulfur in the presence of piperidine, either under conventional heating or microwave or ultrasound irradiation conditions, to yield thienopyridines 52. Again, the use of microwave and ultrasound irradiation requires less time to bring reactions to completion. Microwave reactions were observed to proceed faster than those promoted by ultrasound. The structure of compound 52a was assigned by using X-ray crystallography.

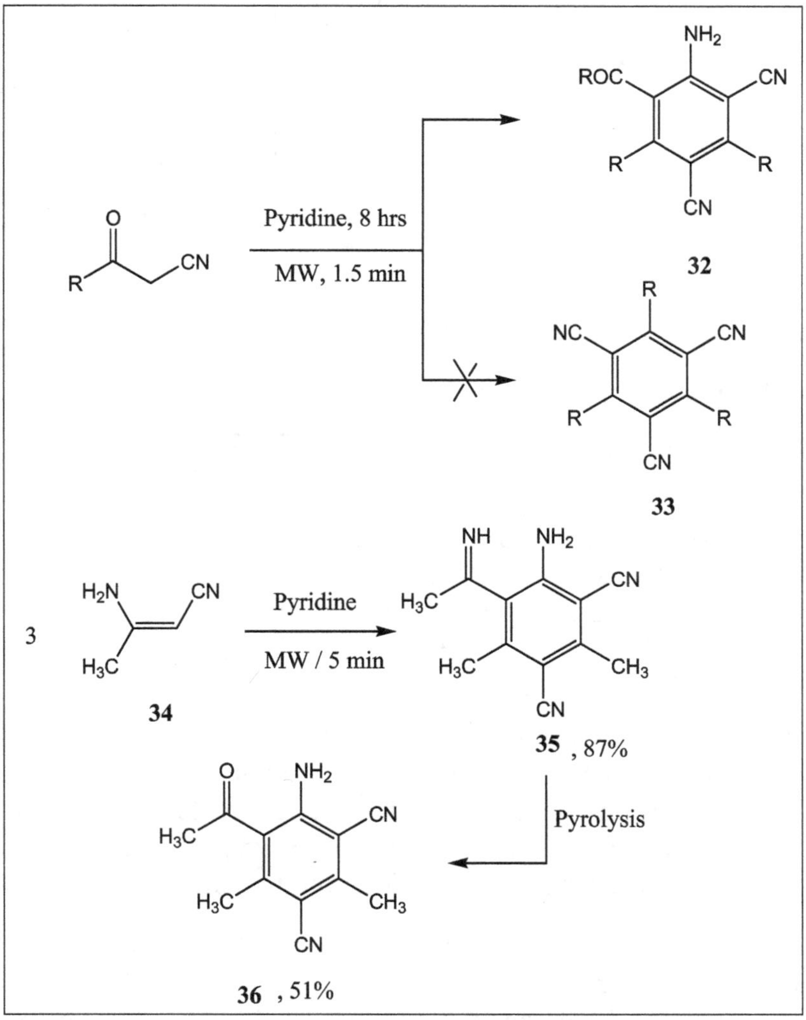

Syntheses of compounds 32ac and 36.

Table: Yields of compounds 32ac.

Compounds	R	Yield%
32a	C_6H_5	85
32b	p-ClC_6H_4	87
32c	CH_3	27

Substituted 2-amino-2-chromenes 56 have received considerable attention owing to their importance as pigments, agrochemicals, and cosmetics.

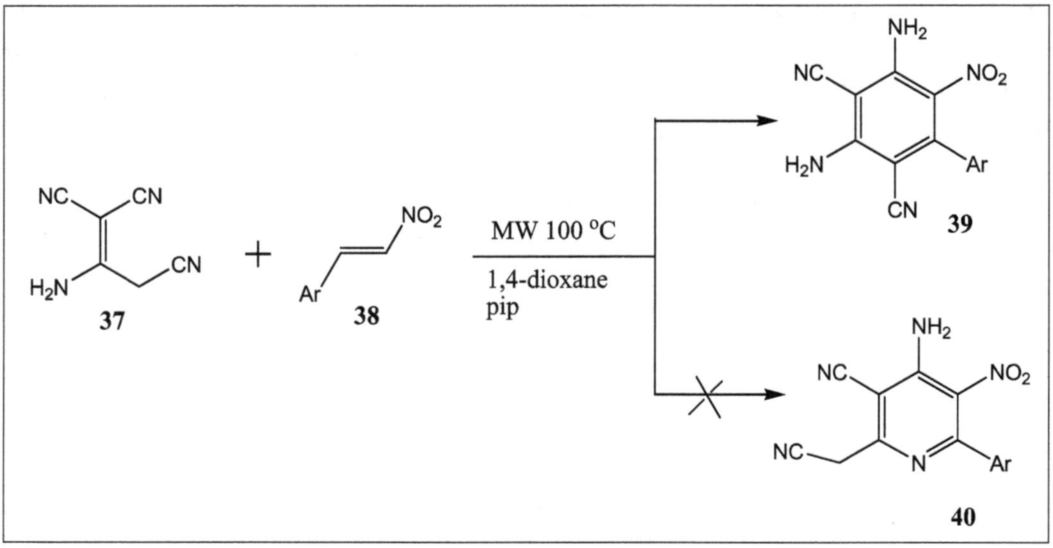

Synthesis of compound 39af.

Table: Yields of compounds 39af.

Compounds	Ar	Yield%
39a	C_6H_5	73
39b	p-ClC_6H_4	70
39c	p-$CH_3OC_6H_4$	72
39d	m-$NO_2C_6H_4$	70
39e	2-Furyl	72
39f	3,4-$Cl_2C_6H_3$	70

The synthesis of 2-amino-2-chromenes using conventional heating techniques requires prolonged times, and products are formed in only moderate yields. In contrast, an efficient, high-yielding threecomponent synthesis of these target molecules involves the utilization of controlled microwave heating of aldehydes with active methylene nitriles and a-naphthol 55 in ethanol containing a catalytic amount of piperidine at 80 8C for 58 min.

We also developed a simple and efficient method for carrying out a one-pot three-component synthesis of the Biginelli 4-aryl-3,4-dihydropyrimidine-2(1H)- ones 58 through reaction of aldehydes, ethyl acetoacetate, and urea or thiourea under controlled microwave heating conditions.

Table: Yields of compounds 52ad.

Compounds	R	Ar	Yield%
52a	Butyl	o-$MeO_2CC_6H_4$	84
52b	Hexyl	o-$MeO_2CC_6H_4$	88
52c	Cyclohexyl	o-$MeO_2CC_6H_4$	89
52d	Benzyl	o-$MeO_2CC_6H_4$	87

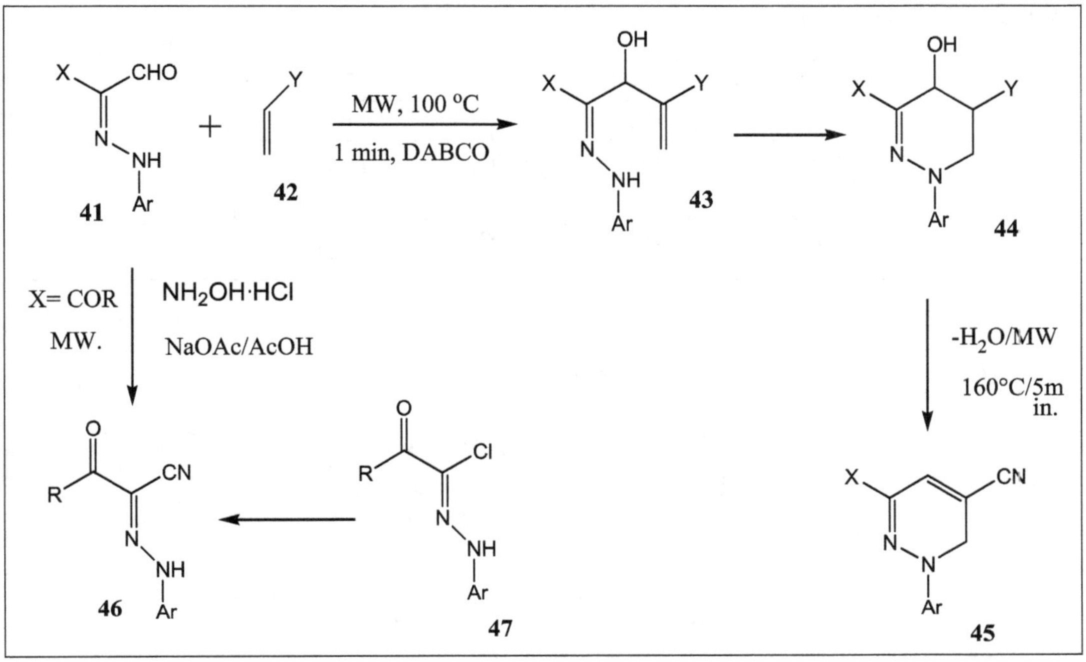

Syntheses of compounds 45 and 46.

Under microwave irradiation conditions, 59 reacts to afford a mixture of 60 and 61, of which the former undergoes self-condensation to yield 62 upon microwave irradiation in an acetic acid solution in the presence of acidic zeolite.

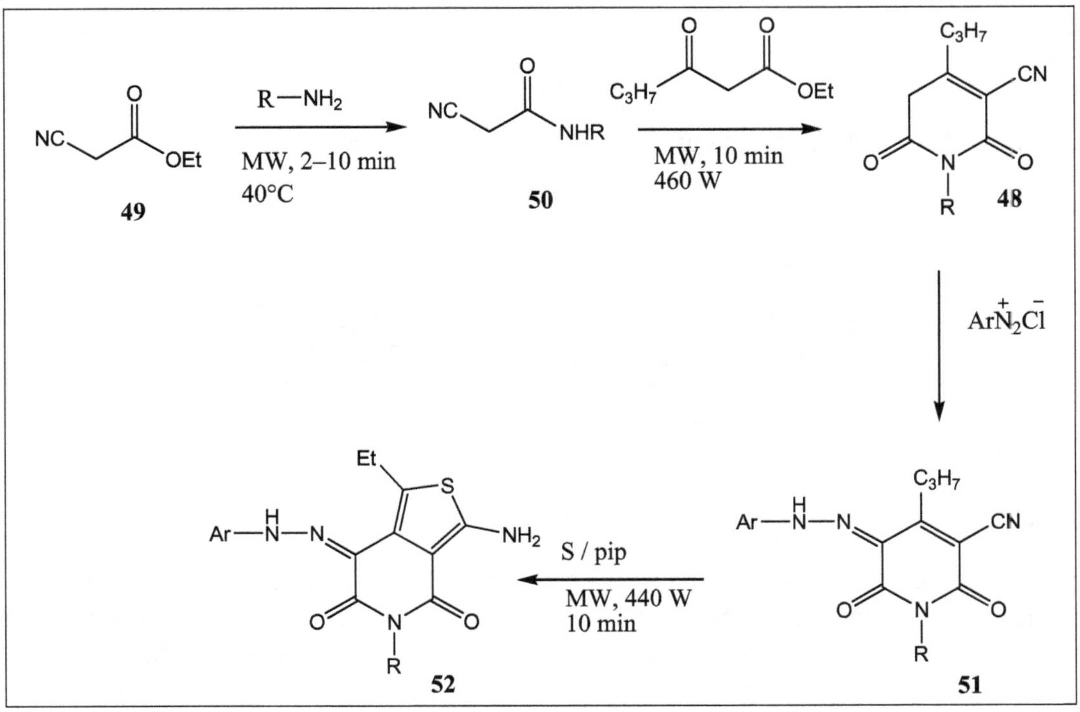

Syntheses of compounds 52ad.

Table: Yields of compounds 56ad.

Compounds	X	Ar	Yield%
56a	CN	C_6H_5	93
56b	CN	p-$CH_3OC_6H_4$	94
56c	CN	p-ClC_6H_4	88
56d	CN	p-$NO_2C_6H_4$	85

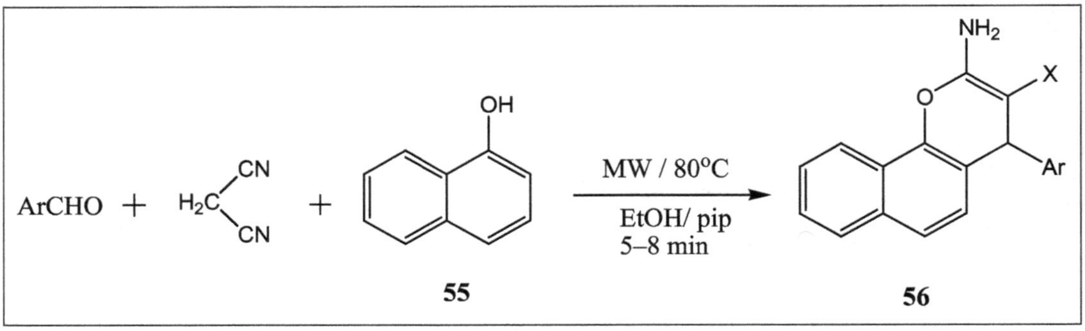

Syntheses of compounds 56a-d.

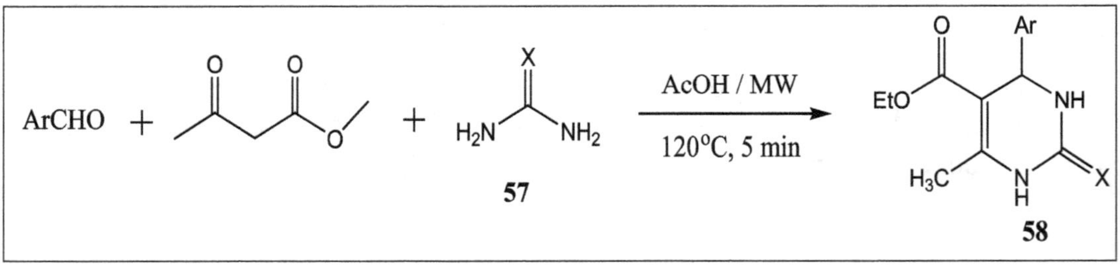

Syntheses of compounds 58-f.

Table: Yields of compounds 58af.

Compounds	X	Ar	Yield
58a	O	C_6H=	90
58b	O	p-$CH_3OC_6H_4$	91
58c	O	p-ClC_6H_4	88
58d	O	p-O_2NC_6H=	80
58e	S	C_6H=	87
58f	S	p-$CH_3OC_6H_4$	88

Solar Thermal Energy as an Energy Source

Alternative and freely available clean energy sources. An important indicator of the greenness of a chemical reaction is found in its energy efficiency (the cost and environmental friendliness of the energy, and its applicability on a large scale). Solar energy is essentially free and, because its use does not require raw materials, no pollutants are produced. Thus, solar energy can be regarded as a clean reagent.

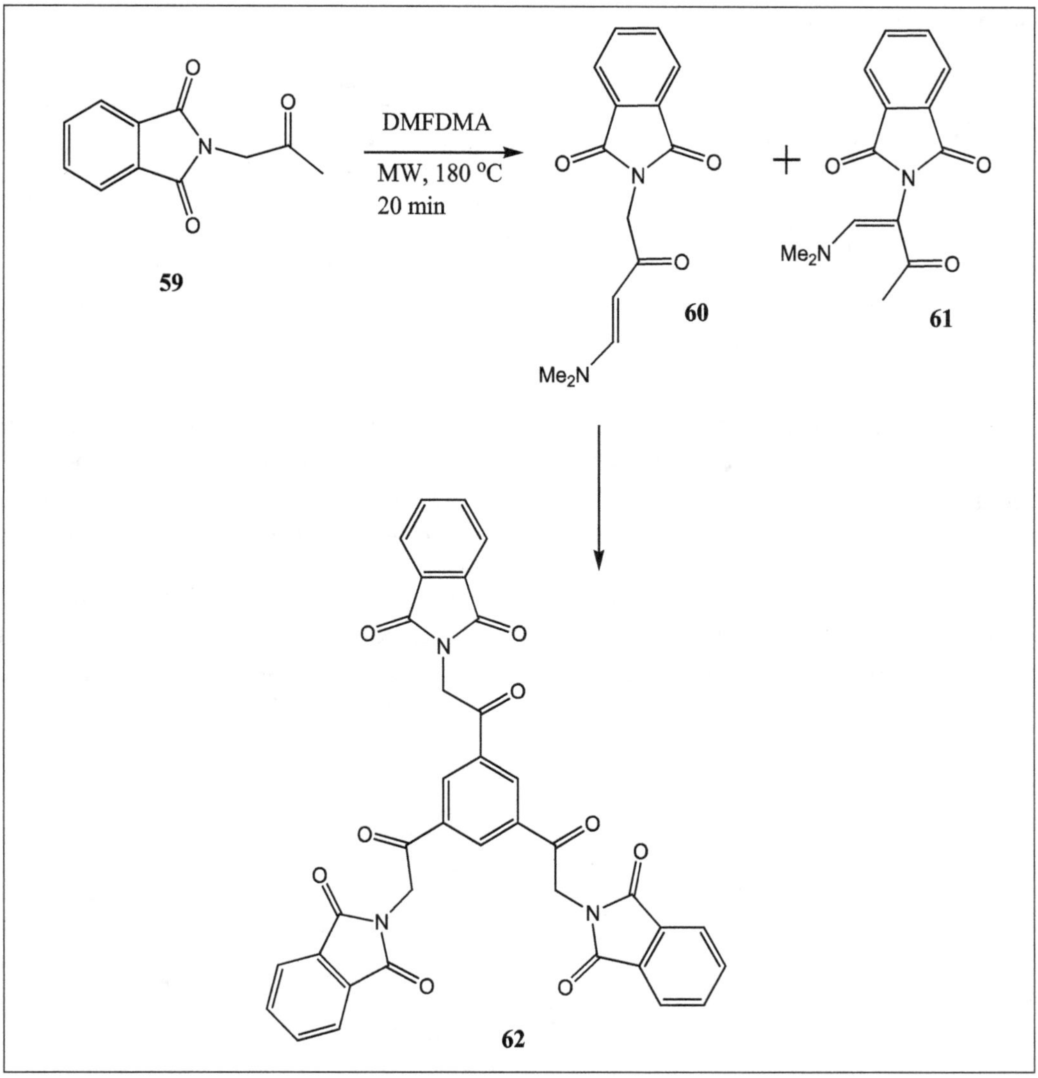

Syntheses of compound 62.

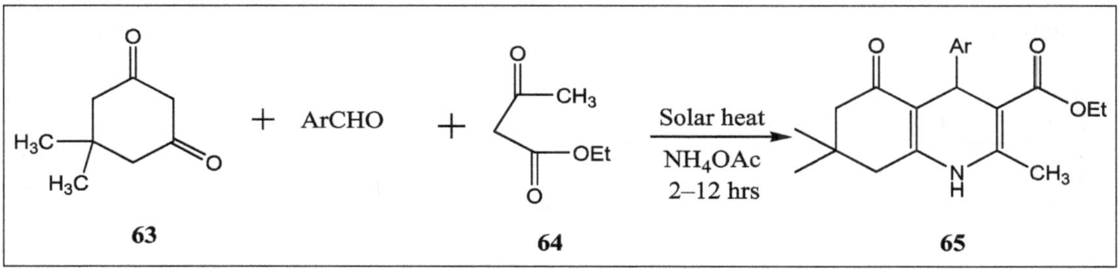

Syntheses of compounds 65a-g.

A description of our first utilization of solar thermochemical energy in organic synthesis was published in 2008. The process involved a four-component synthesis of polyhydroquinoline derivative 65 via the reaction of dimedone 63, aromatic aldehydes, ethyl acetoacetate 64, and ammonium acetate employing solar heating.

Table: Yields of compounds 65a-g.

Compounds	Ar	Yield%
65a	C_6H_5	92
65b	p-$CH_3OC_6H_4$	92
65c	1-Naphthyl	88
65d	p-ClC_6H_4	87
65e	2-Furyl	90
65f	p-$CH_3C_6H_4$	88
65g	m-$O_2NC_6H_4$	83

Green conditions have been developed for the synthesis of substituted 2-aminothiophenes 70, employing multicomponent reactions of a ketone 66 with active methylene nitrile and elemental sulfur induced by solar thermal energy. The procedure proved to be efficient and simple in terms of conducting the reaction and isolating products.

The tetra-substituted imidazole scaffold is a core component of many biological systems as well as several natural products. A number of synthetic approaches for the construction of this scaffold have been published, but all involve the use of high temperatures, expensive metal precursors, and catalysts that may be harmful to the environment.

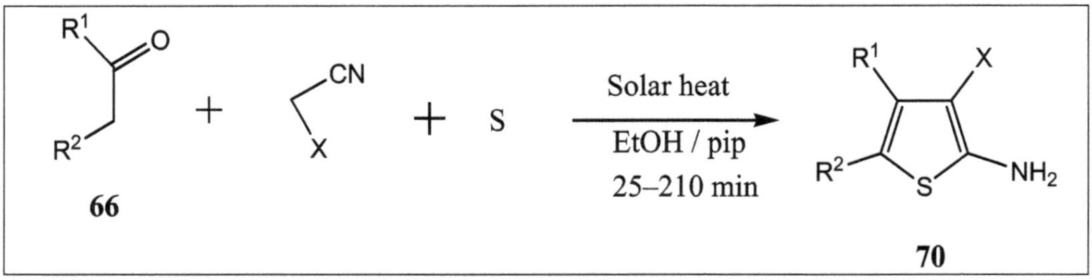

Syntheses of compounds 70a-f.

Table: Yields of compounds 70a-f.

Compounds	R1	R2	X	Yield%
70a	CH_3	$CO_2C_2H_5$	CN	72
70b	CH_3	$CO_2C_2H_5$	$CO_2C_2H_5$	70
70c	Cyclohexy1	CH_3	CN	88
70d	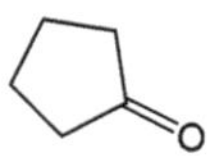	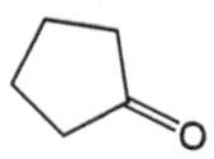	$CO_2C_2H_5$	75
70e	CH_3	CH_3	CN	73
70f	H_3C-N 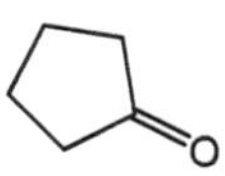	H_3C-N 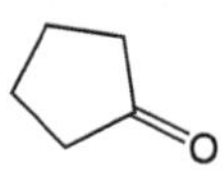	$CO_2C_2H_5$	75

We observed that exposing a solution of benzoin 71, aromatic aldehydes, aromatic amines 72, and ammonium acetate in dichloromethane in the presence of catalytic amount of high surface area SiO_2 to direct sunlight for 2-25 h leads to the formation of the corresponding tetra-substituted imidazoles 73 in excellent yields.

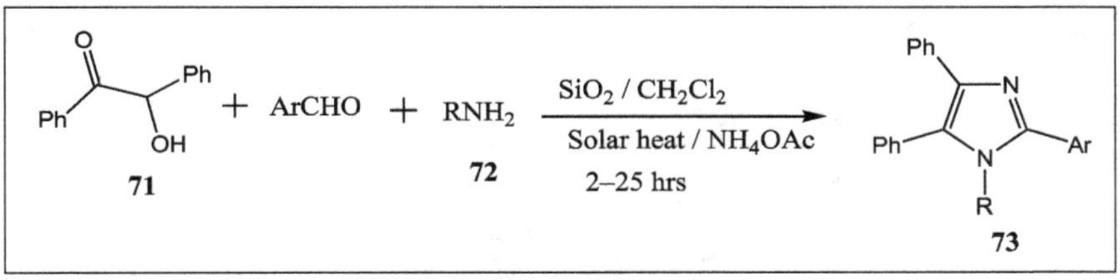

Syntheses of compounds 73a-f.

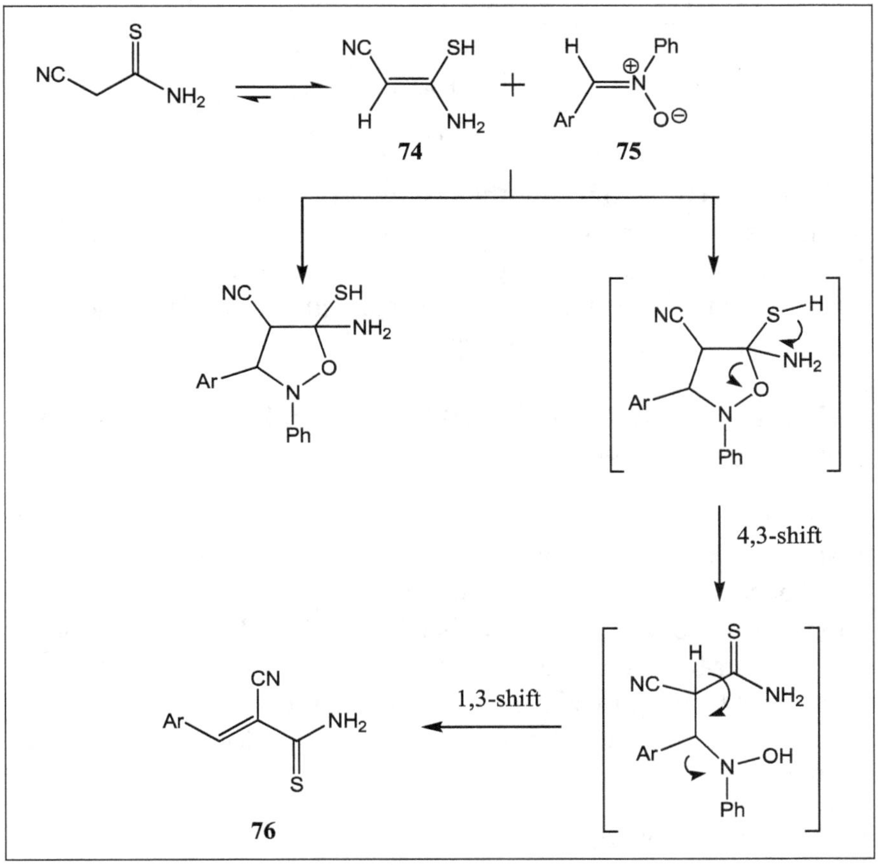

Syntheses of compounds 76a-f.

Table: Yields of compounds 73a-f.

Compounds	Ar	R	Yield%
73a	C_6H_5	$C_6H=$	88
73b	C_6H_5	$C_6H_5CH_2$	80
73c	p-ClC_6H_4	C_6H_5	93

73d	p-$CH_3C_6H_4$	C_6H_5	90
73e	m-$O_2NC_6H_4$	p-$CH_3C_6H_4$	89
73f	p-ClC_6H_4	p-ClC_6H_4	84

Table: Yields of compounds 76a-f.

Compounds	Ar	Yield%
76a	$C_6H=$	63
76b	p-$CH_3OC_6H_4$	73
76c	p-$O_2NC_6H_4$	69
76d	o-$O_2NC_6H_4$	65
76e	p-Me_2NCH_4	66
76f	2-Thienyl	54
76g	2-Furyl	52

We, observed that reactions of cyanothioacetamide 74 with nitrones 75 in EtOH under solar thermal energy conditions afford 76 in good yields. The reaction of 74 with diphenylnitrone furnished the corresponding 2-cyano-3-phenylprop-2-ene 76a. It is worth mentioning that 76a has not previously isolated as a solid product. A mechanism that accounts for the formation of 76 is presented in Scheme 19.

Utility of Heterogeneous Acid and Base Catalysis to Replace Homogenous Alternatives

We extensively investigated the reactivity of functionally substituted cinnamonitriles 77 toward active methylene compounds, electron rich aromatics, azolones, and alkyl heteroaromatic carbonitriles in the presence of homogeneous base catalysts, usually piperidine or pyridine. This effort led to the preparation of a variety of pyrans 78, thiopyrans 79, benzopyrans 80, naphthopyrans 81, pyranoazoles 82, thiazolopyridines 83, phthaloazinones 84, and cinnolines 85. Several of the compounds synthesized in this effort showed interesting biological activities, and described in patents.

Table: Yields of compounds 80, 81 and 83.

Compounds	Ar	X or Y	Yields%
80	$C_6H=$	CN	74
81	C_6H_5	CN	76
83	C_6H_5	CN	96

As a consequence, synthetic approaches to these types of compounds have attracted much attention recently.

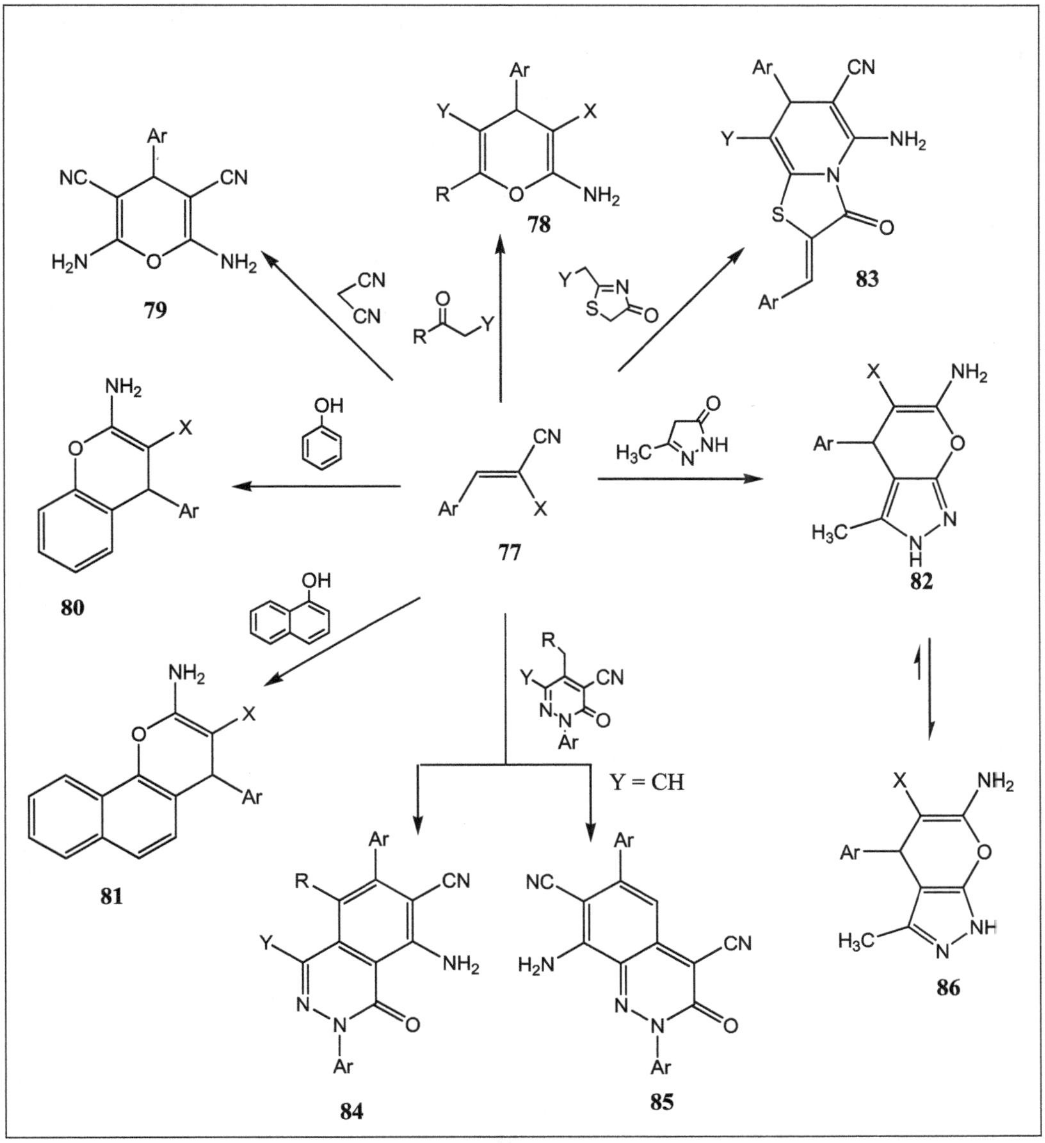

Syntheses of compounds 78-86.

Table: Yields of compounds 90a-c.

Compounds	R	Yield%
90a	C_6H_5	72
90b	2-Furyl	75
90c	2-Thinyl	72

In 2008, we reported on utility of chitosan, a naturally occurring biopolymer obtained by treating chitin with strong alkali, as a heterogeneous catalyst for these reactions. It occurred to us that it might be valuable to probe in detail the actual structures of some of these products, which have been questioned. Chitosan is hydrophilic basic catalyst and, as such, it has been used successfully as a catalyst for Michael addition reactions.

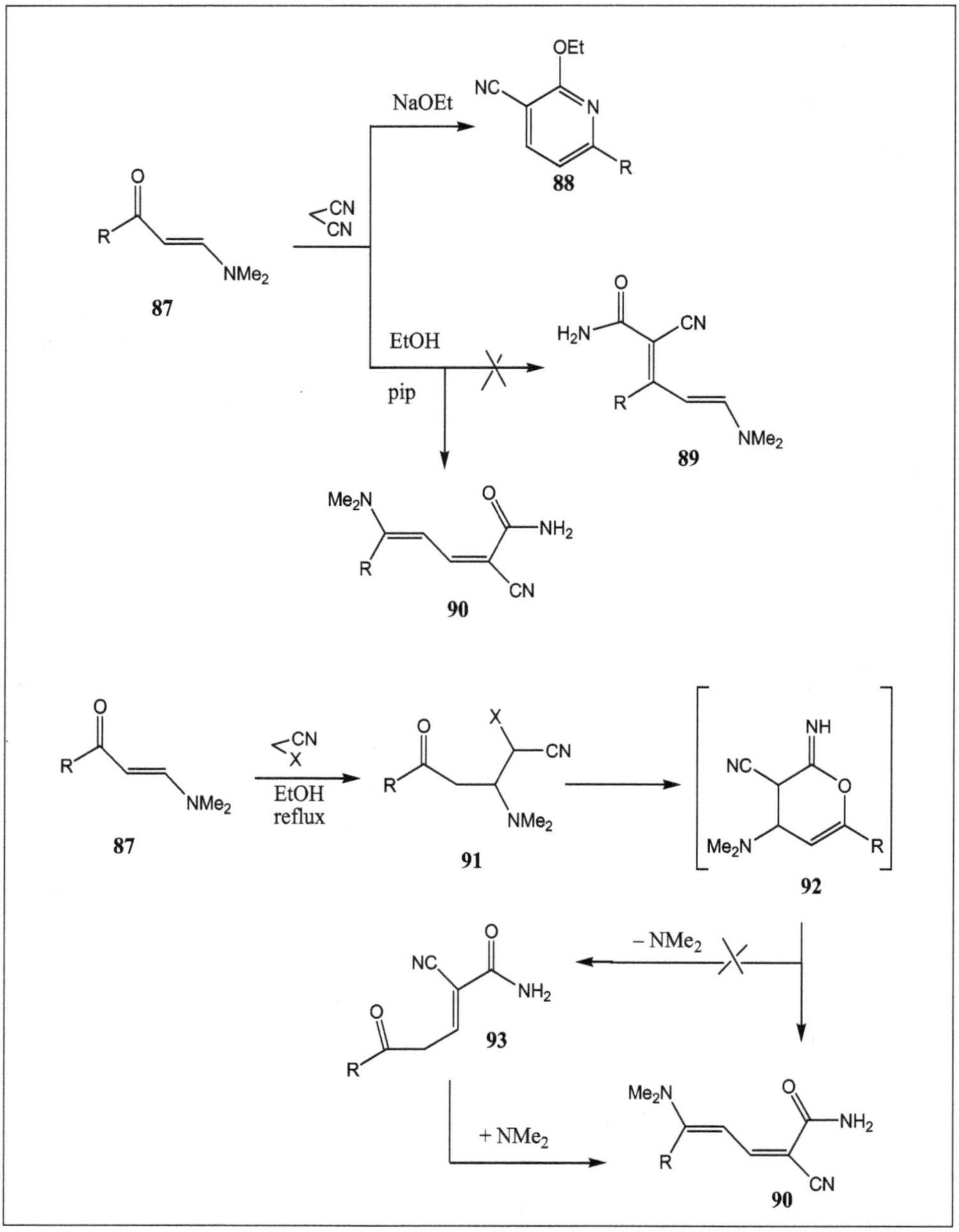

Syntheses of compounds 90a-c.

It was observed that reactions of 77 (XCN) with acetoacetic ester and malononitrile in the presence of chitosan afford 78 and 79, respectively, in almost the same yields described for reactions in the presence of piperidine.

Similarly, reactions of this substance with phenols, naphthols, and pyrazolone afforded the respective products 80, 81, and 82. The latter product was incorrectly assigned as 86 earlier. The reaction of 77 with thiazolylacetonitriles and ethyl thiazolylacetate afforded 83 as expected despite claims to the contrary. Pyridazinylcarbonitriles (YCO-2Et) afforded 84 whereas when YCN 85 was formed. All of the observed yields are similar to those given in the original reports.

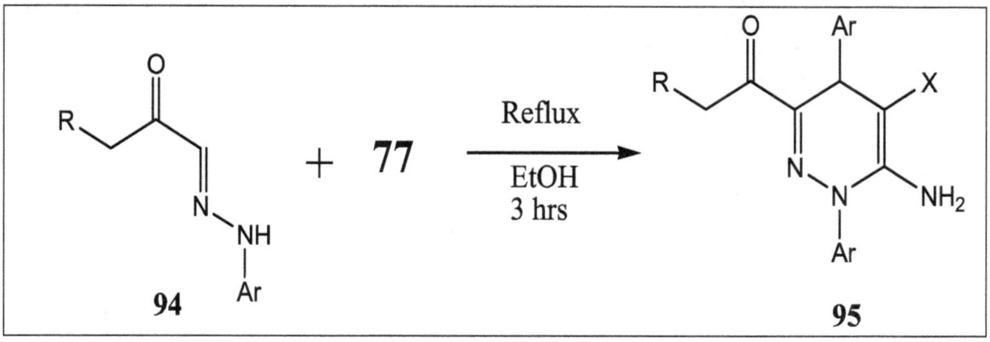

Syntheses of compounds 95a-d.

We have also utilized chitosan as base catalyst to promote addition reactions of malononitrile and ethyl cyanoacetate to form the enaminone 87. Initially, AlOmran et al. reported that 88 is the product of the reaction of these substances promoted by ethanolic sodium ethoxide, but suggested that 89 is produced when the reaction occurred in ethanolic piperidine. Utilizing both 15N NMR, HMBC NMR methods, and a subsequent X-ray crystal structural analysis, we demonstrated that the reaction product is really 90.

Table: Yields of compounds 95a-d.

Compounds	Ar	R	X	Yield%
95a	C_6H_5	H	CN	84
95b	C_6H_5	CH_3	CN	82
95c	m-$O_2NC_6H_4$	H	CN	85
95d	p-ClC_6H_4	H	CN	80

87 reacts not only with malononitrile but also with ethyl cyanoacetate and benzoylacetonitrile to yield dienamides, which are believed to be formed via intermediates 91, 92, and 93. The possibility that formation of 92 takes place, which then reacts further by elimination of dimethylamine, was excluded based on experimental observations.

Chitosan has also been used as catalyst for the addition of 94 to 77 yielding 95, thus replacing piperidine reported in the initial paper as the catalyst of choice.

Eco-friendly, solid Lewis acid catalysts as well as ionic liquids have also been employed as replacement for their noneco-friendly counterparts. In 2007 we reported the utilization of 97, prepared by the reaction of indole 96 with cyanoacetic acid/acetic anhydride mixture following the procedure reported by Slatt et al. as a building block for the synthesis of substituted indoles. This approach appeared to us to be an interesting route to 3-oxoalkanomitriles. However, less electron-rich aromatics failed to undergo the reaction employing the conditions developed. We envisioned that In Cl_3 would act as an efficient catalyst for this condensation and, indeed, this catalyst did promote the reaction of 98 to form oxoalanonitriles 99. Also, 101 and 102 have been produced from 100 in 70% yields. Abdel-Khalik and Elnagdi reported that enaminones 103a are converted into 1,2,5-triaroylbenzenes 104a in refluxing in acetic acid.

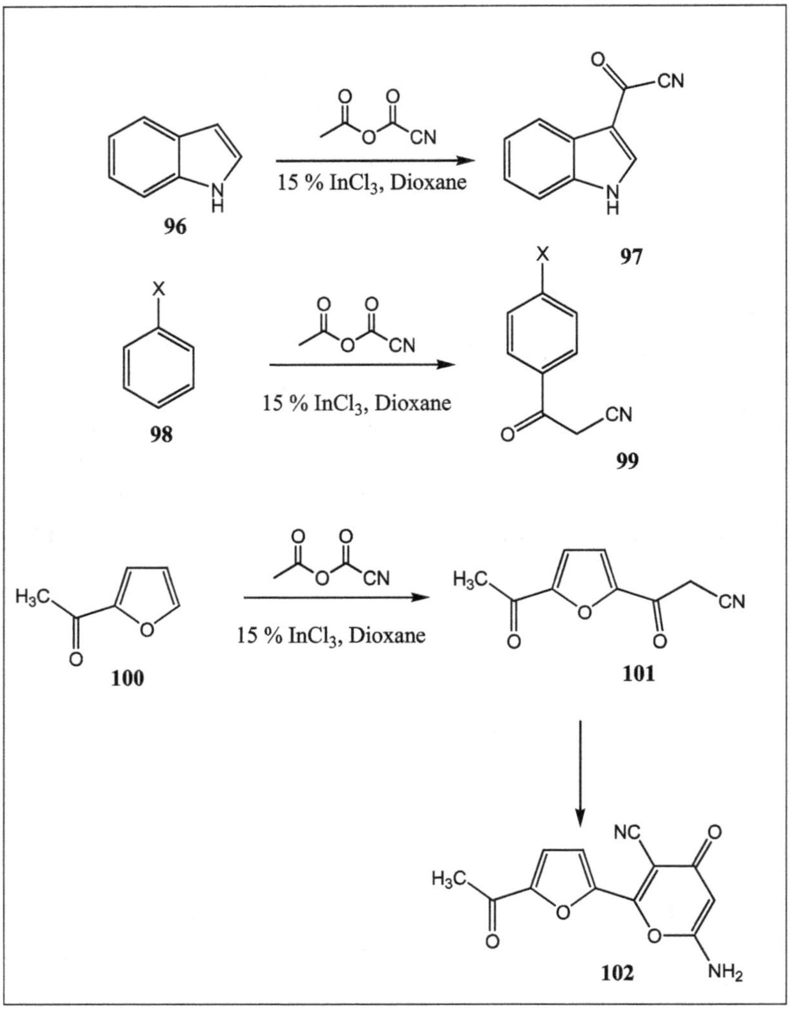

Syntheses of compounds 97, 99, 101, and 102.

Subsequently Makhseed et al. extended this approach to the synthesis of benzene-1,3,5-trials 104 (RH) and triethylbenzene-1,3,5-tricarboxylates 104 (ROH). Recently Al-Zaydi et al. were able to affect the conversion of 103d to 104d more efficiently using pyridine hydrochloride as an ionic liquid and MW as the energy source. Although 103e in refluxing acetic acid reacts to generate 105 only, when the process is carried out in the presence of pyridinium hydrochloride 104e is formed. Very recently, Al-Mousawi et al. showed that these conversions can be affected by fusing enaminones in the presence of montmorillonite K10.

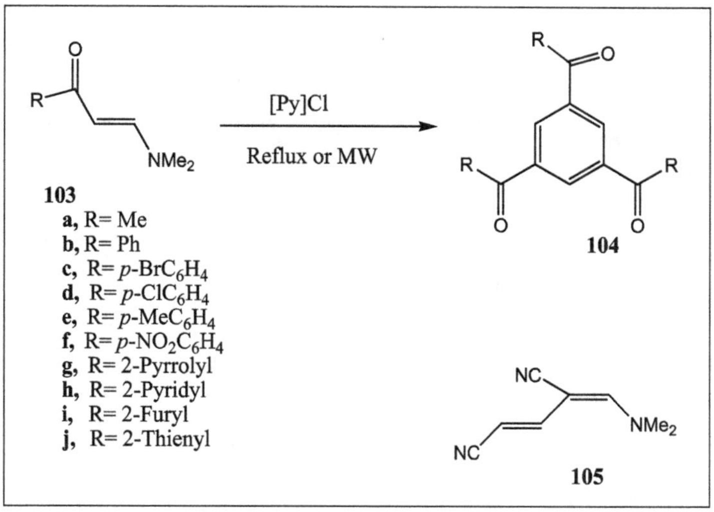

Synthesis of compound 105

Moreover, these workers provided evidence that the reaction takes place in a stepwise manner. Specifically, the results of crossover experiments showed that mixtures of 106 and 107 afford both 108 and 109. In addition, the reaction of 110 with ethyl propiolate was shown to afford a mixture of 111113. On heating with dimethyl acetylenedicarboxylate in the presence of montmorillonite K10, 110 is converted to 114.

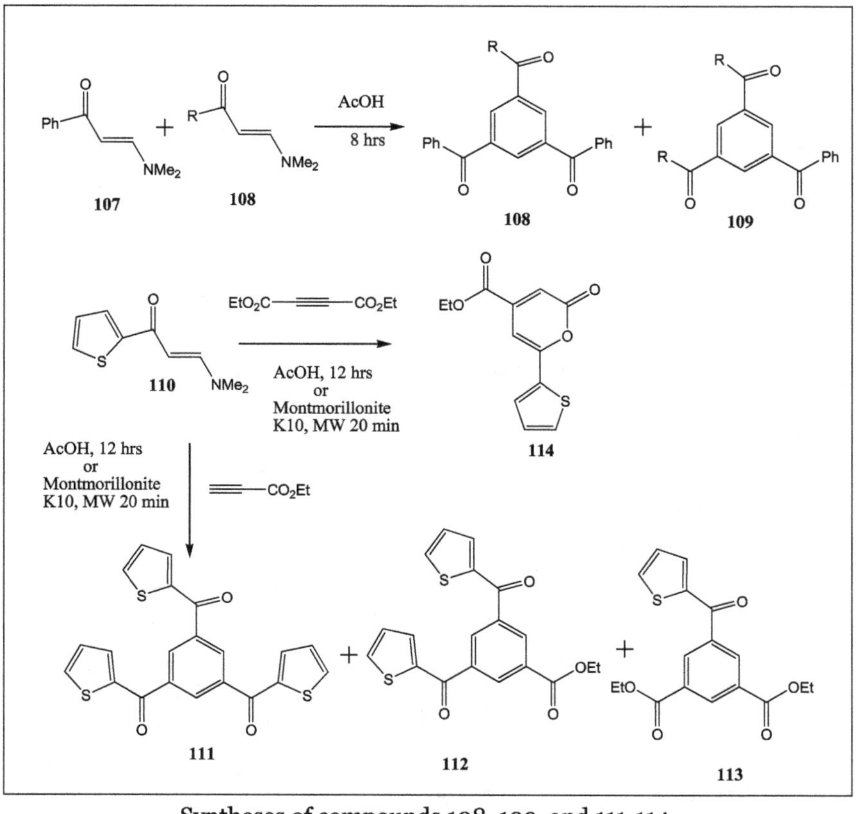

Syntheses of compounds 108, 109, and 111-114.

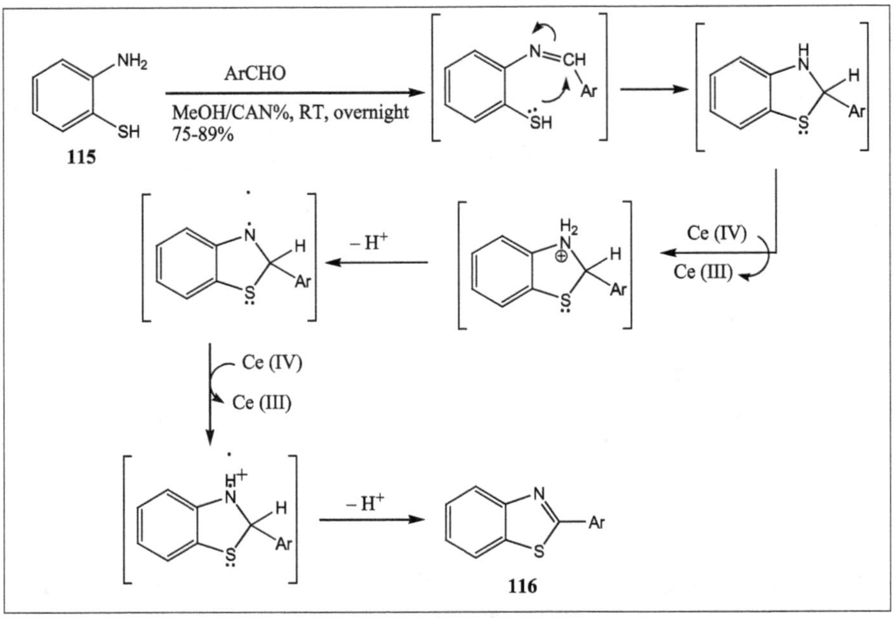

Syntheses of compounds 116a-h.

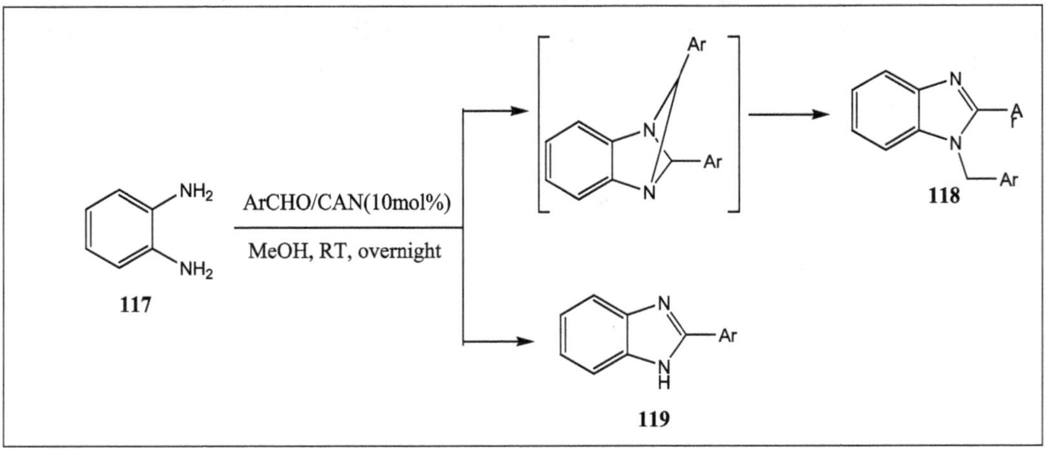

Syntheses of compounds 118a-c and 119a-b.

Table: Yields of compounds 116a-h.

Compounds	Ar	Yields%
116a	$C_6H=$	75
116b	p-$CH_3OC_6H_4$	78
116c	1,4-$C_6H_9(OCH_3)_2$	77
116d	m-$O_2NC_6H_4$	88
116e	o-ClC_6H_4	89
116f	o-$CH_3OC_6H_4$	80
116g	p-$O_2NC_6H_4$	87
116h	2-Thienyl	79

Ceric (IV) ammonium nitrate (CAN) is a convenient and widely used reagent for affecting a broad range of reactions owing to advantages that include solubility in water and various organic solvents, inexpensiveness, eco-friendly nature, uncomplicated handling, fast conversions, and convenient work up procedure. We have initiated a program aimed at exploring the potential of CAN in organic transformation. This effort led to the development of a onepot CAN-catalyzed synthesis of 2-arylthiazoloes 116 from the reaction of 2-aminothiophenol 115 and aromatic aldehydes.

The reactions of o-phenylenediamine 117 with aromatic aldehydes in MeOH at room temperature, catalyzed by Cerium (IV) ammonium nitrate (CAN), afford either 118 and/or 119. The results clearly show that this process proceeds mainly via two different routes. The first involves 1:1 condensation followed by oxidation to afford 2-aryl-1-substituted-1H-benzimidazoles. The second process is a 1:2 condensation that generates 2-aryl-1-arylmethyl-1H-benzimidazoles. Thus, in these reactions both of the expected products are obtained in ratios that depend on nature of the solvent. For example, in methanol both 118 and 119 are obtained.

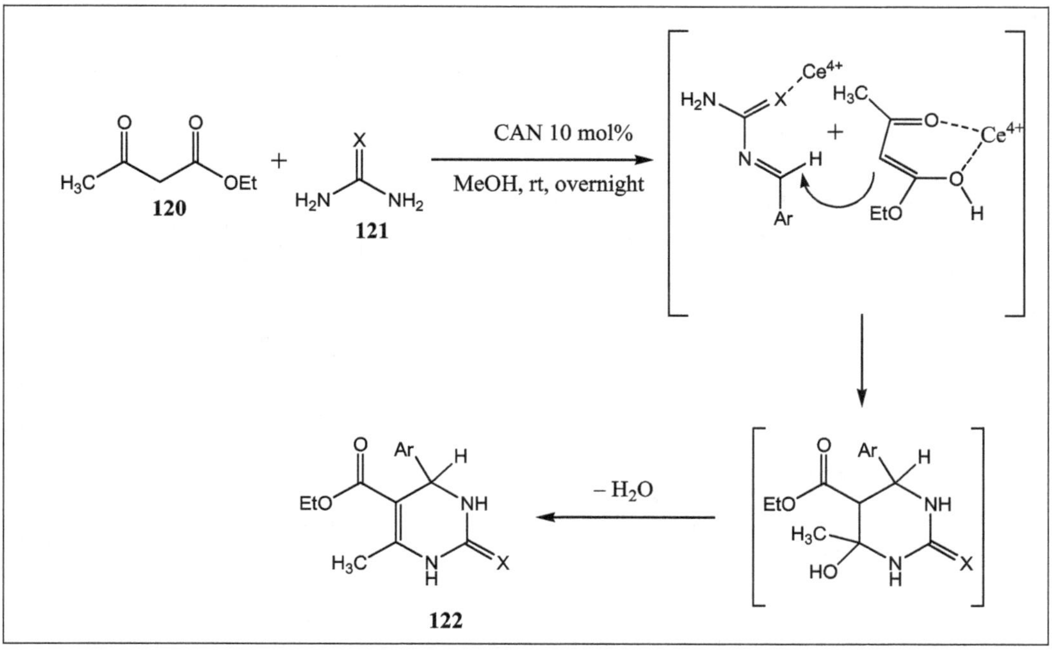

Syntheses of compounds 122a-g.

Table: Yields of compounds 118a-c and 119a-b.

Compounds	Ar	Yield%
118a	p-$CH_3OC_6H_4$	90
118b	7,4-C_6H_3-OCH_3	89
118c	m-$O_2NC_6H_4$	15
119a	o-$O_2NC_6H_4$	55
119b	p-ClC_6H_4	70

Although extensively utilized as a one-electron oxidant, CAN's use as a Lewis acid catalyst in CN bond forming reactions leading to heterocyclic compounds is somehow limited . We recently observed that a simple and highly efficient procedure for conducting the Biginelli condensation reaction of aldehydes, b-ketoesters 120, urea, or thiourea 121a,b at ambient temperature involves the use of CAN as a Lewis-acid catalyst. A mechanism to account for the formation of the products.

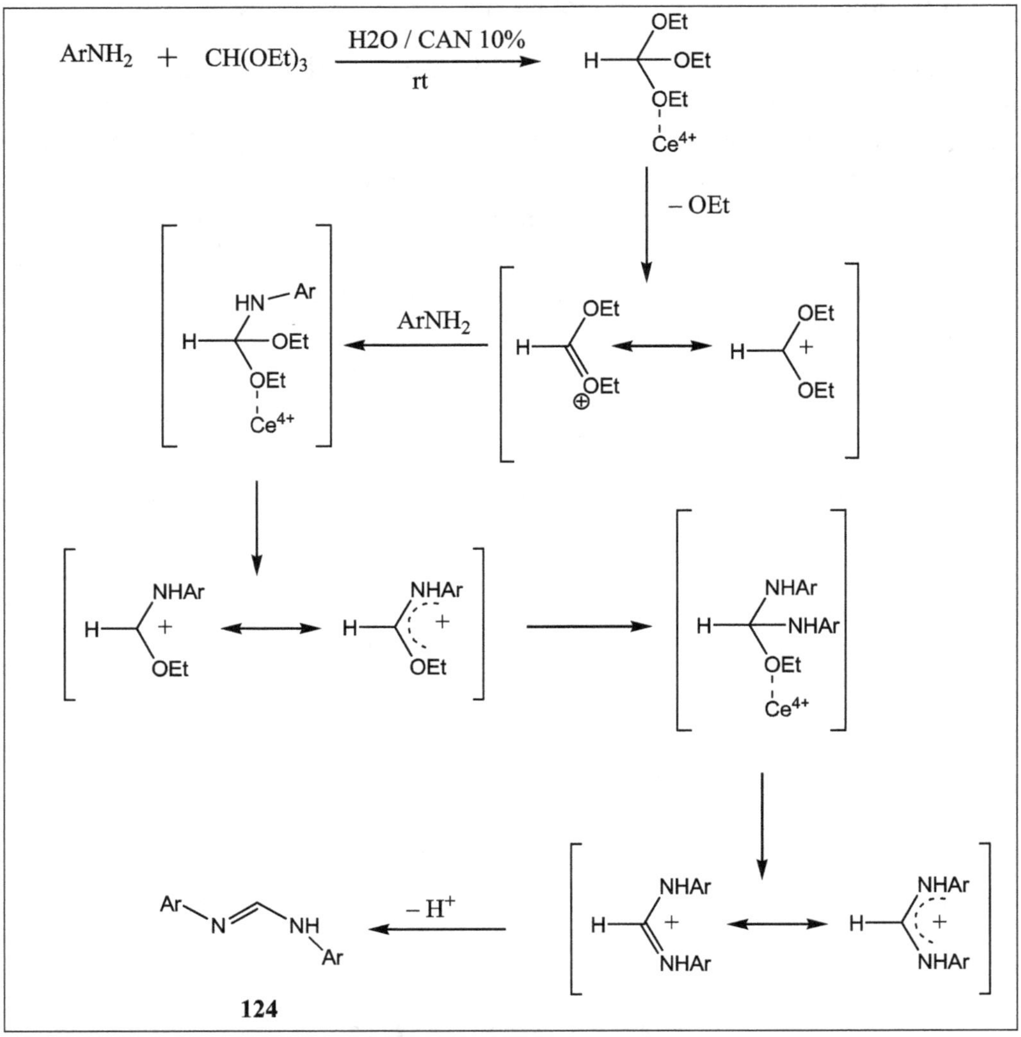

Syntheses of compounds 124a-h.

Table: Yields of compounds 122a-g.

Compounds	X	Ar	Yields%
122a	O	C_6H_5	90
122b	O	p-CH_3OC_6H4	93
122c	O	m-$O_2NC_6H_4$	96
122d	O	p-ClC_6H_4	88

122e	S	C_6H_5	94
122f	S	p-$CH_3OC_6H_4$	92
122g	S	p-$O_2NC_6H_4$	93

N,N-diaryl-formamidines are of considerable interest in fields related to organic and medicinal chemistry. Exchange reactions of N,N-dimethylformamidines or acetamidines by a variety of amines have been reported to be the most popular method for the synthesis of symmetrical and unsymmetrical formamidines. We have developed a simple and green route for the synthesis of N,N-diaryl-formamidines 124 via the reaction of aromatic amines with triethylorthoformate 123 in water at room temperature catalyzed by CAN acting as a Lewis acid catalyst.

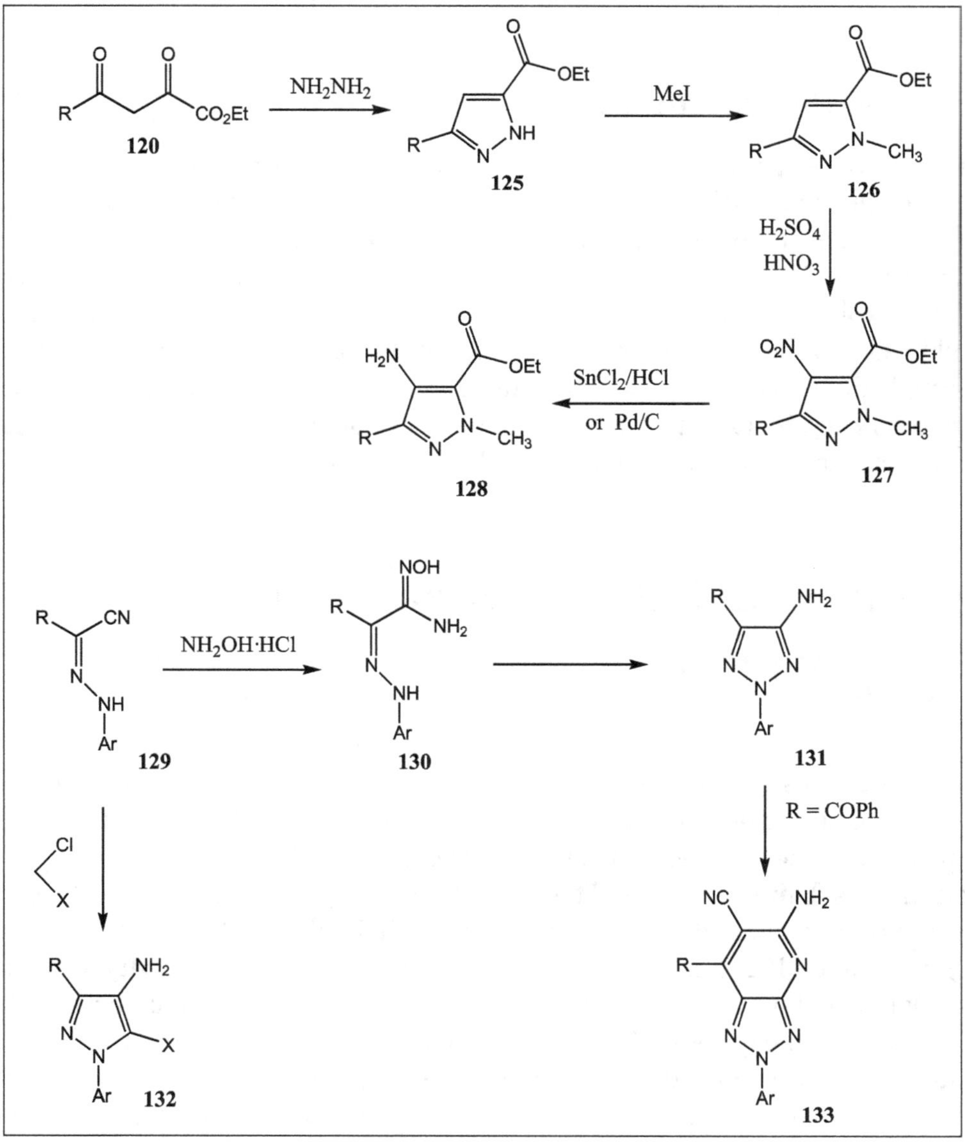

Syntheses of compounds 128, 132, and 133.

Table: Yields of compounds 124a-h.

Compounds	Ar	Yield%
124a	C_6H_5	93
124b	p-$CH_3C_6H_4$	94
124c	p-$CH_3OC_6H_4$	95
124d	p-$O_2NC_6H_4$	86
124e	o-ClC_6H_4	88
124f	p-ClC_6H_4	87
124g	p-BrC_6H_4	88
124h	5-Methyl-3-yl	88
	Pyrazole	88

Greener Synthetic Approaches

4-Aminopyrazoles-5-carboxylic acid derivatives are interesting intermediates in the synthesis of pharmaceuticals. Moreover, reported synthetic approaches to these substances are rather inefficient hazardous and/ or nonconcise. For example, 128, a precursor of Viagra, is prepared using a multistep sequence that includes production of potentially explosive nitropyrazoles by employing a hazardous nitration reaction mixture followed by a heavy metal reduction. Recently we described a direct synthesis of 132 through the reaction of 129 with functionally substituted alkyl halides. Similarly 129 was converted to a Zaperinsate analog via reaction with NH4OH to yield intermediate 130 that then reacts with malononitrile to yield 133.

Synthesis and Characterization of Nanocomplexes

A convenient synthesis of a series of La(III) complexes (in nanoscales) with different organic ligands was prepared by green chemical method. The physicochemical properties of the nanocomplexes were obtained. The SEM and TEM techniques were used to confirm the nanosize and morphology. The cytotoxicity of the newly synthesized compounds was demonstrated and tested for their anticancer activity against common pathogenic organisms and against human epithelial colorectal adenocarcinoma cells ($Caco_2$), Vero, and MCF-7 cells. By examining its pharmacological activity, we were able to identify new potent and selective anticancer agents. Also, the structures succeeded to affect and stop the growth of Klebsiella pneumonia strain, and the results revealed promising activity. In addition, an electrospinning technique was used for the fabrication of novel bio-medicated nanofibers which are applied for wound dressing as local chemotherapy for skin injuries.

Studies on metal-based drugs offer promise in the fight against cancer, which is the main aim of the inorganic chemistry.

The metal-based anticancer drugs are driven by the necessity to fill the gap in tumor chemotherapy in order to minimize the undesirable side effects and enlarge the spectrum of activity for more tumor types and metastatic cancers. The Schiff base structures play a key role in studying the mechanism of the transformation and racemization reaction in biological systems due to the similarity of their structures to the natural biological system.

Among various metal complexes, La(III) complexes have been intensively investigated due to their more physiological activities and lower toxicities after coordination with ligand, so, in recent years people have paid great interest to synthesis, DNA interaction, and anticancer activity of La(III) complexes.

Dressings play multimodal roles in wound healing process, such as preventing infections, absorbing exudates, and maintaining moisture, and thus are indispensable. Electrospinning is the method to fabricate ultrafine fibers from polymer solution(s) or melt(s) with diameters ranging from nanometers to micrometers.

Experimental

Organic solvents such as ethyl alcohol and dimethylformamide (DMF) were spectroscopically pure from the British Drug House (BDH). In addition to the 5-bromosalicylaldehyde, $La(NO_3)\cdot 6H_2O$ (Merck) were also used.

Instrumentation

In open capillaries melting points (°C, uncorrected) were determined on a Gallenkamp melting point apparatus (Sanyo Gallenkamp, Southborough, UK). In the Microanalytical Center, Cairo University (C, H, N, and S) were carried out. ^{1}H NMR spectra (DMSO-d_6) were measured at Bruker FT-400MHZ spectrophotometer. The IR spectra were recorded on a Perkin Elmer 437 IR spectrophotometer (400–4000 cm^{-1}) (KBr technique). Thermogravimetric analysis was performed using a Shimadzu TGA-50H with a flow rate of 20 ml min^{-1}. The UV-vis spectra were recorded on a Perkin Elmer Lambda 3B UV-vis spectrophotometer. X-ray diffraction were recorded at room temperature (~25°C) on Empyrean X-ray diffractometer equipped. The patterns were run with Cu target (CuKα radiation), and the tube operated at 45 kV and 30 mA. The size and morphology of the nanocomplexes were characterized with a scanning electron microscope (SEM) with gold coating and TEM.

Cytotoxicity Assay

Different concentrations of the nanometal complexes were tested for their cytotoxicity against Vero, Caco2, and MCF-7 cell lines using (MTT) Thiazolyl Blue Tetrazolium

Bromide method. In brief, Vero, Caco2, and MCF-7 cells (10 × 10^3) were cultured in a 96-well plate overnight at 37°C, 5% CO_2, and 88% humidity. The total volume of used DEMEM supplemented medium and the synthesized compound supernatant was 200 µL with final concentrations of 10, 20, 30, 40, 60, 80, and 100 mg/L. The plate was incubated at 37°C and 5% CO_2 for 3 days. After incubation, debris and dead cells were removed by washing three times with fresh medium. Twenty microliters of MTT solution (5 mg/mL of MTT in PBS buffer) was added to each well and shook for 5 min at 150 rpm to thoroughly mix the MTT into the media. The cells were incubated at 37°C and 5% CO_2 for 3–5 h to metabolize MTT by viable cells. Two hundred microliter dimethyl sulfoxide (DMSO) was added to each well and shook again for 5 min at 150 rpm, and then the viability of the cells was calculated by measuring the optical density at 630 nm subtracted from optical density at 570 nm. The percentage of viability cells was calculated by comparison with control cells (without adding the synthesized compounds to the cells) using the equation (A) test/(A) control × 100.

Flow Cytometry

To discriminate living cells from dead cells or for cell cycle analysis, propidium iodide (PI) can be used. This analysis is based on the stoichiometric binding of PI to intracellular DNA. At the end, cells were washed with PBS and collected by trypsinization. Cells were then resuspended in warm PBS and fixed with 4 ml ice-cold ethanol. Finally, in darkness, cells were stained with 0.5 ml of warm PI solution; 7 ml of PI solution consists of 0.35 ml of PI stock solution (1 mg/ml), 0.7 ml RNase A stock solution (1 mg/ml), and 6 ml of PBS and incubated in darkness for 30 minutes. The samples were kept on ice until flow cytometric analysis.

Preparation of (Chitosan-La(III) Nanocomplex) Composite Nanofibers

Chitosan (CS) was dissolved in 50% acetic acids at room temperature under moderate stirring for 12 h to form a homogeneous solution, and then the solution of La complex nanoparticles was quietly adding. The total solution was fixed to 3 wt.%. The solution was filled in a 20 ml NORM-JECT Luer Lock tip plastic syringe having an 18-gauge stainless steel needle with 90° blunt end. The electrospinning setup included a high voltage power supply, and a nanofiber collector of aluminum foil that covered a laboratory-produced roller with the diameter of 10 cm. The collector was placed at 10 cm tip-to-collector distance (TCD).

Synthesis

Synthesis of Complex Nanoparticles by Sonochemistry Method (Green Chemistry)

Ten milliliter of 0.1 M solution of $La(NO_3)_3.6H_2O$ in ethanol was positioned in a high-density ultrasonic probe, operating at 24 kHz with a maximum power output of 400 W. Ten milliliter of 0.1 M solution of N'-((E)-5-bromo-2-hydroxybenzylidene)-6-(((E)-5-bromo-2-hydroxybenzylidene)amino)-4-oxo-2-thioxo-1,2,3,4-tetrahydrothieno[3,2-d]pyrimidine-7-carbohydrazide H_2L^1, N-(2-mercaptophenyl) benzamide H_2L^2, and 2-((2-mercapto- phenyl) imino)-1,2-diphenylethan-1-ol H_2L^3 was added dropwise. The obtained precipitates were allowed to evaporate at room temperature to obtain the complex nanoparticle in powder form.

Elemental Analyses for Nanometal Complexes

The analytical data of the prepared metal complexes suggest the structures as given in Figure. The physicochemical results of the newly synthesized compounds were presented in Table. The obtained results were in good agreement with those calculated for the suggested formulae. The melting points were sharp, indicating the purity of the prepared nanometal complexes. The proposed formula of the nanometal complexes is according to the following general equations:

Table: Elemental analysis and some physical measurements of La(III) nanocomplexes (1-3). *ohm $^{-1}$cm^{2}mol^{-1}.

Compd. code	(Empirical formula)	M.P. (°C)	Color (yield %)	(%) found (calcd.)				
				C	H	N	S	Λm*
1	$[(La)(L1)(NO_3)(H_2O)]$	<350	Faint orange (80.54)	29.90 (30.01)	1.32 (1.56)	9.92 (10.00)	7.58 (7.62)	20.51
2	$[La(H_2L_2).(NO_3)_2(H_2O)_2]NO_3.H_2O$	110	Dark gray (85.49)	2.94 (3.06)	27.75 (27.87)	5.40 (5.71)	9.80 (10.00)	57.88
3	$[La(H_2L_3)(NO_3)3.H_2O]$	165	Page (79.80)	2.83 (2.90)	36.54 (36.31)	4.55 (4.83)	8.33 (8.47)	3.39

^{1}H NMR Spectra

The ^{1}HNMR spectra of the La (III) nanocomplex 1 showed a singlet peak at 8.00 ppm assigned to azomethine proton, in addition to the multiplet signals due to aromatic protons. It was found that the spectra of complex 2 display singlet peaks due to SH and NH groups at δ3.5 and 5.4 ppm indicating the involvement of these groups in coordination with the metal atoms without deprotonation.

Table: ^{1}H NMR data for La(III) nanocomplexes.

Compd. no.	δ_{ph}	$\delta_{CH=N}$	δ_{SH}	δ_{NH}	δ_{OH}
1	6.39–8.46	8.00	–	–	–
2	6.99–7.48	–	3.5	5.4	–
3	7.00–8.00	–	3.47	–	10.4

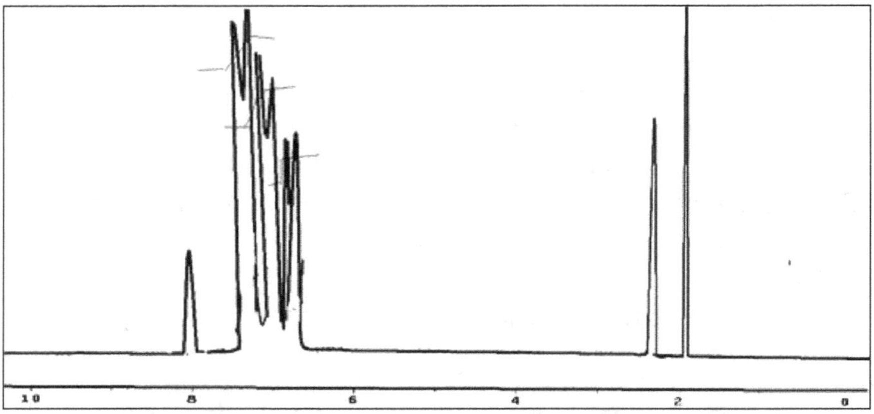

1H NMR data for La(III) diamagnetic complex (1).

In addition to the ^{1}H NMR spectra of complex 3, showed singlet peaks due to the O—H proton at 10.4 ppm and the S—H proton peak at 3.47 ppm. This analysis showed that the O—H and S—H groups share in the complexation a loss of their protons. Moreover, the spectra of the complexes showed multiplet signals at 6.99–7.48 ppm attributed to the aromatic protons for all complexes.

IR Spectra and Mode of Bonding

In order to ascertain the mode of bonding, the IR spectrum of the free ligands was compared with those of their La complexes. By careful comparison of the spectra of the complexes with those of the free ligands, it was found that the band at 3430 cm^{-1} is found to be disappeared by complexation suggesting the involvement of OH groups of the phenolic groups in complex formation with their deprotonation. The bands due to NH and SH are shifted to higher or lower frequencies indicating the participation of –N and –S donors in the coordination, while the bands due to C=O are unchanged. The strong bands at 1622 and 1298 cm^{-1} due to $\upsilon_{C=N}$ and $\upsilon_{C—O}$ (azomethine and phenolic groups) are shifted to lower wave number 1609–1565 and 1280–1242 cm^{-1}, respectively, in metal complexes indicating the participation of nitrogen of azomethine and oxygen of phenolic group in complexation. The complexes showing also three bands at the range 1480–1400, 1380–1350, and 1170–1020 cm^{-1} indicate the unidentate coordination mode of the nitrate group. The curves also, showed that shift in the peaks of O—H, C—O, C—S groups (complex 3) due to their sharing in the complexation and that recorded by the M—O, and M—S peaks. All these results are in good agreement with the conductance data. Another evidence for the coordination is the observance of new bands in the far IR region at 570–505, 520–440, and 432–524 cm^{-1} which may be due to $\upsilon_{M—O}$ $\upsilon_{M—N}$ and $\upsilon_{M—S}$, respectively, that are not observed in the spectrum of the free ligands.

Molar Conductivity Measurements

The conductivity Λ_m value of the La nanocomplexes 1–3 can be calculated by using the relation Λ_m = K/C. The complexes were (10^{-3} M) dissolved in DMF, and the molar conductivities of their solutions at 25±2°C were measured. It is concluded from the results that complexes 1 and 3 are found to have molar conductance values of 20.51 and 3.39 ohm^{-1} mol^{-1} cm^2, respectively, indicating that these complexes are nonelectrolytic and monomeric in nature. Also the values indicate the bonding of the nitrate ions to metal cations. The molar conductivity value of complex 2 is seen to be 57.88 ohm^{-1} mol^{-1} cm^2 indicating the ionic nature of this complex, and the nitrate ion is outside the coordination sphere.

Electronic Spectral Data

The assignments of the significant electronic spectral absorption band of the nanometal complexes are given in Table. Two absorption peaks were observed in the spectra range of 210–227 and 279–337 nm due to π-π* and n-π* transitions, respectively, due to benzene and the azomethine (CH=N) function. Moreover, the spectra of complexes indicate no significant absorption in the visible region due to the absence of f–f transition, since f–f transitions are Laporte-forbidden and very weak in nature.

Table: Magnetic moment and electronic spectral data of La(III) nanocomplexes.

Compd. no.	μ_{eff} (B.M.)	Absorption bands (nm)	
		π-π*	n-π*
1	Diamagnetic	269	337
2	Diamagnetic	210	280
3	Diamagnetic	227	279

Magnetic Susceptibility

The following equation was used to calculate the magnetic susceptibility of the La nanocomplexes:

$$X_g = \frac{C(bal)(1)(R-R^o)}{10^{-9} m}$$

where C, calibration constant of standard; l, length of sample in filled tube; R^o, reading of empty sample tube; R, reading with sample; χ_g gram magnetic susceptibility.

The gram magnetic susceptibility of the complexes was found in the range -2.2098×10^{-6} $-(-2.3013 \times 10^{-6})$ cm^3 mol^{-1}, respectively. Negative values of χ_g show that these complexes are diamagnetic in nature or there is no significant magnetic moment.

Thermal Analysis

Thermogravimetric analyses (TGA) of the La nanocomplexes were used to determine the thermal stability of the complexes, decide whether the water molecules are lattice or coordinated, and finally suggest a scheme for thermal decomposition of these chelates. Heating rates were suitably controlled at 10°C min^{-1} under nitrogen atmosphere, and the weight loss was measured from the ambient temperature up to ~ 600°C. The weight loss for each chelate was calculated within the corresponding temperature ranges.

Also, TGA curve of the La(III) complex of nanoscale $[(La)(L^1)(NO_3)(H_2O)]$ (1) showed three thermal stages. The first stage represents the loss of H_2O molecule of coordination, HNO_3 and ½O_2 with a found weight loss of 11.72% (cacld. 11.54%) within the temperature range of 130–180°C; the subsequent second step (180–300°C) corresponds to the loss of $C_5H_3BrN_2O$ with an estimated mass loss of 22.60% (calcd. 22.27%). The final stage at 300–520°C corresponds to loss of rest of the organic part ($C_{12}H_7BrN_3O_2S_2$) with an estimated weight loss 44.25% (calcd. 43.94%) leaving 4 C and La metal as a metallic residue. The overall weight loss amounts is 78.57% (calcd. 77.75%).

Powder X-ray Diffraction Studies

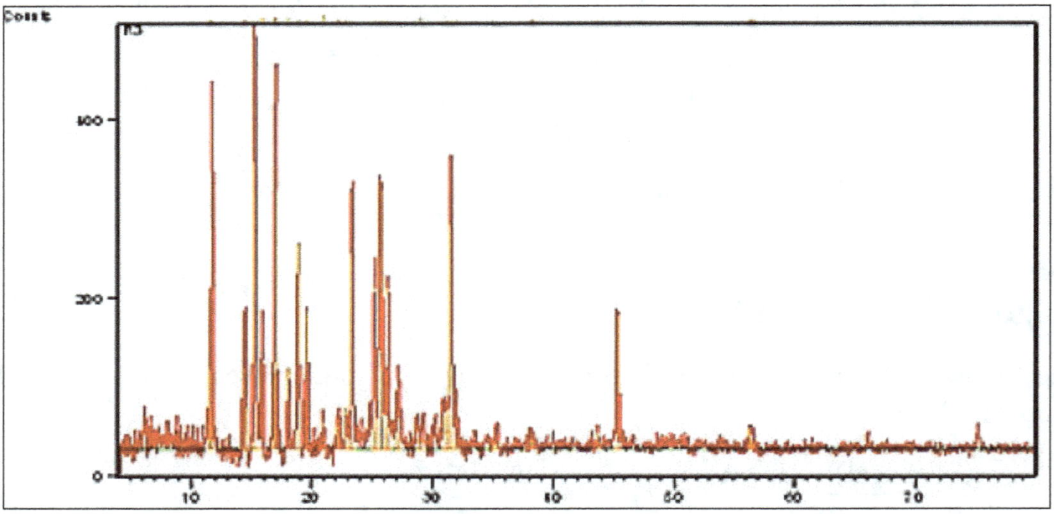

X-ray powder diffractogram of complex (1) nanoparticle.

The diffraction of X-rays from the planes of a crystal (diffraction analysis) is considered as one of the important methods of the analytical work. By using X-ray diffraction which depends on the crystal properties of solids, we can identify the crystal structure of various solid compounds and identify the actual compounds from its structure, and we can also determine the arrangement of molecules in crystal. X-ray diffractometry is an important technique as it is a nondestructive, non-contrast, fast, and sensitive one. Obtained XRD data of H_2L^1 ligand and its La(III) complex 1 of nanoparticle (Figure) show that XRD pattern of the ligand is different in comparison with the XRD pattern of the nanocomplex. The obtained data indicate that the complex is more crystalline in nature than that of the ligand which has an amorphous phase. The average particle size of the complex (1) can be estimated from Debye-Scherrer equation:

$$L = K\lambda / \beta \cos\theta$$

where β the breadth of the observed diffraction line at its half intensity maximum; K, the shape factor, 0.89; λ, the wavelength of the X-ray source used in XRD; θ, the corresponding incidence angle.

The average particle size of the nanoparticles is 15 nm.

EDX Spectra

The amount of elements present in the percentage level of the metal complexes was identified by EDX data. EDX spectra are used to calculate the percentage level of the elements present in the metal complexes like C, O, N, S, and La that are present in the La(IIl) nanocomplex 2, shown in figure. The revealed data are in good agreement with that of the elemental analysis.

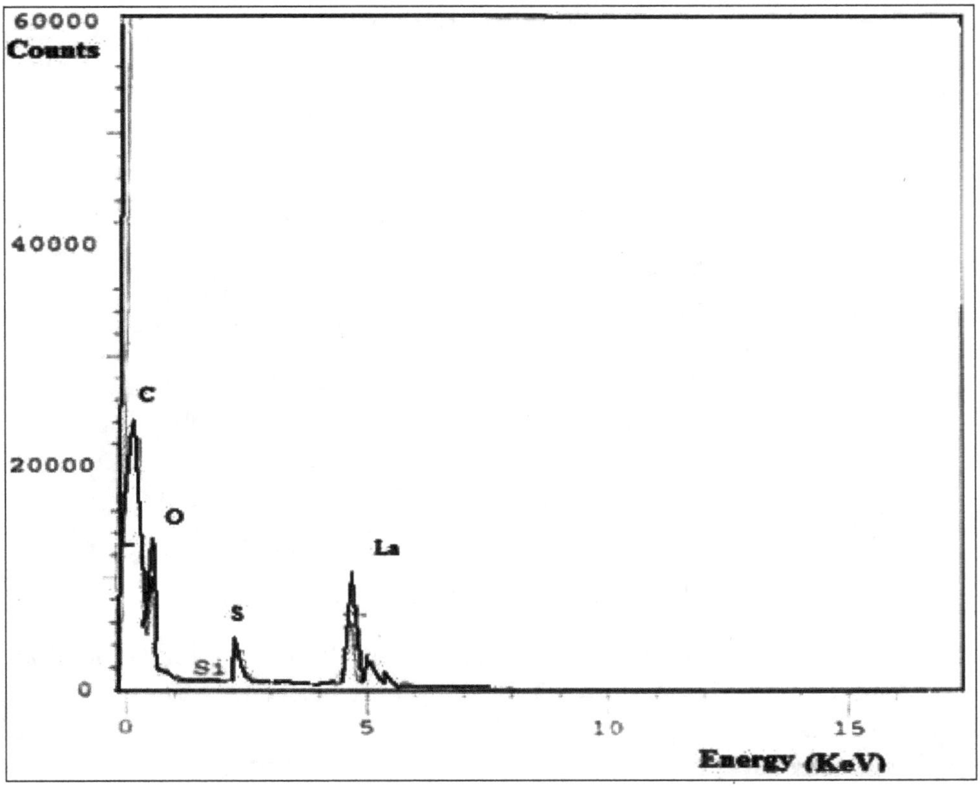

EDX spectra of nano-La(III) complex 2.

SEM Analysis

SEM technique was used to study the morphology and size of nanoparticles. It seems that the particles are semispherical with some agglomerations of the particles. The SEM images further revealed the stabilization of La(III) 1 nanoparticles due to interaction with the Schiff base ligand; this stabilization facilitates penetration of tumor cell membrane and causes the destruction of tumor cell.

SEM images of nanoparticles as produced by ultrasound $[(La)(L1)(NO_3)(H_2O)]$ (1).

TEM Analysis

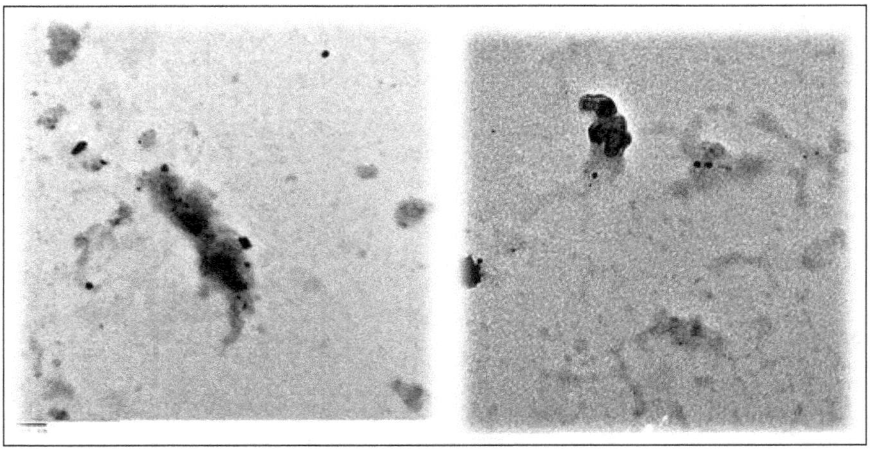

TEM image of La(III) nanocomplex 2.

Transmission electron microscopy (TEM) can be used to directly image nanoparticles at scales approaching a single atom. TEM analysis is performed to examine the size and shape of the nanoparticles. The La(III) complex 2 nanoparticle was fairly uniform in size, spherical in shape, and with the average diameter ranging from 3.2 to 44.9 nm. Electron microscopy analysis allowed confirming visually the observed stability of the obtained nanocomplex.

The fully optimized geometries of the ligand and its metal complexes were shown in figure.

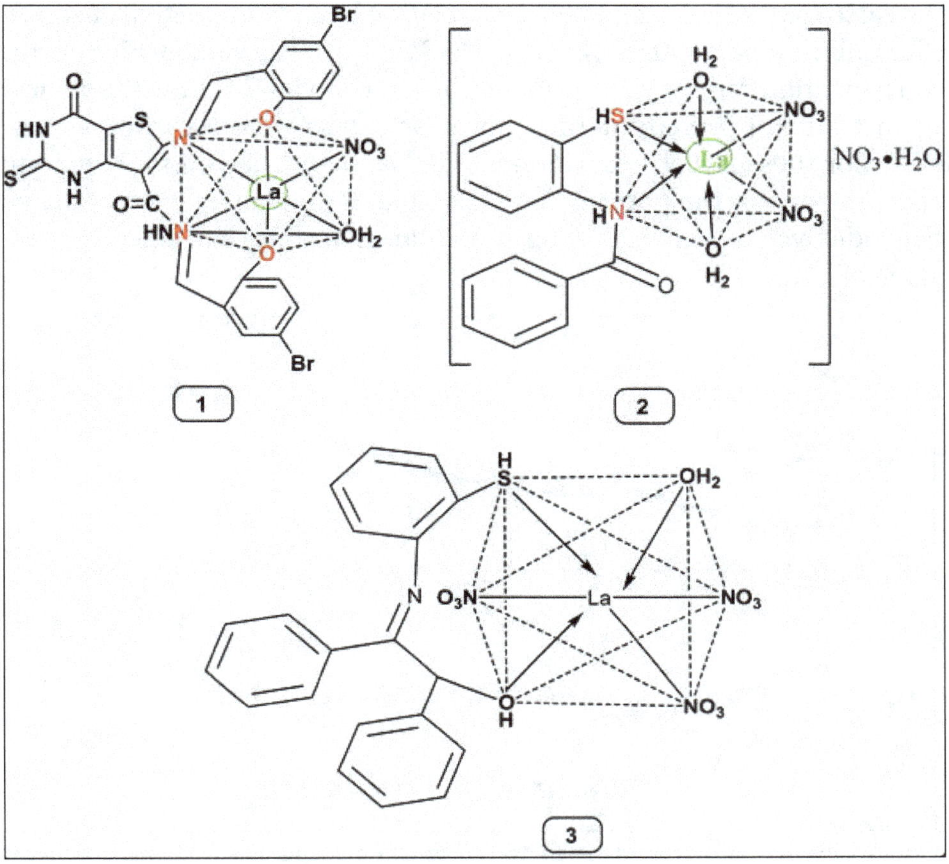

Suggested structure of metal complexes 1–3.

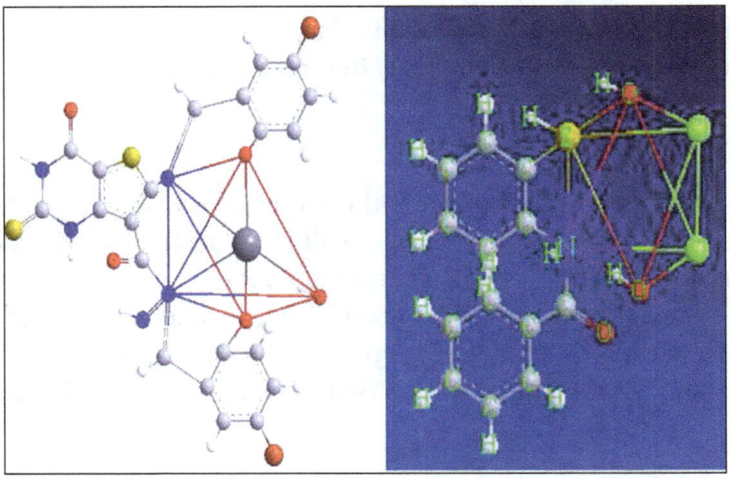

Geometry-optimized structures of (Left) complex 1(Right) La complex 2.

Cytotoxicity and Anticancer Activity

The cytotoxicity of the prepared nanocomplexes has been demonstrated and tested for their anticancer activity against Caco2, Vero, and MCF-7 cell lines. The obtained results revealed that complex 1 had the best recorded anticancer activity against Caco2 cells with inhibition percentage of 72.13. The IC_{50} (μg) (the value which corresponds to the concentration required for 50% inhibition cell viability) for the noncytotoxic doses against human primary dermal fibroblasts cells were recorded as 2.482 μg/ml. The IC_{50} for these structures were recorded as 0.3772 μg/ml for compound 1 on Caco2 cells. In spite of the potency of the sample as anticancer agents, the superior selectivity index values against Caco2 cells than noncancerous cells were recorded with a value of 46.6.

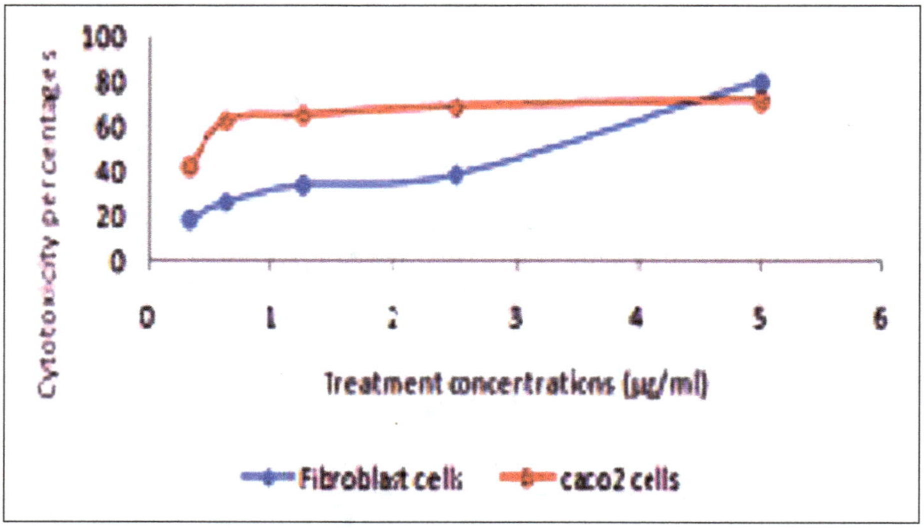

Cytotoxicity and anticancer activity of the prepared compounds against human primary dermal fibroblasts and $Caco_2$ cell lines, respectively.

On the other hand, the other tested compounds have shown moderate activities against all cells; the anticancer activity is less than that recorded for 1.

Flow Cytometry

Flow cytometric analyses of propidium iodide-stained nuclei cells were performed to investigate whether treatment could induce cell cycle perturbations in human epithelial colorectal adenocarcinoma. Cell cycle parameters were compared for Caco2 cells that had been incubated for 24 h with 1 ml of IC_{50} concentrations of sample 1 with untreated (control) cells. As shown in figure, following 24 h incubation with treatment, there was an increase in the proportion of cells in the G2/M phase (51.10%, for treatment 1).

Also, an additional peak was observed, consistent with aneuploid cells which contain more than 4n DNA. This peak may represent a population of cells that has escaped mitotic arrest and continued to replicate as multinuclear cells without dividing (i.e., endoreduplication), or it may be due to a small fraction of tetraploid cells which are already present in the culture, which undergo G2/M arrest, resulting in cells with 8n DNA.

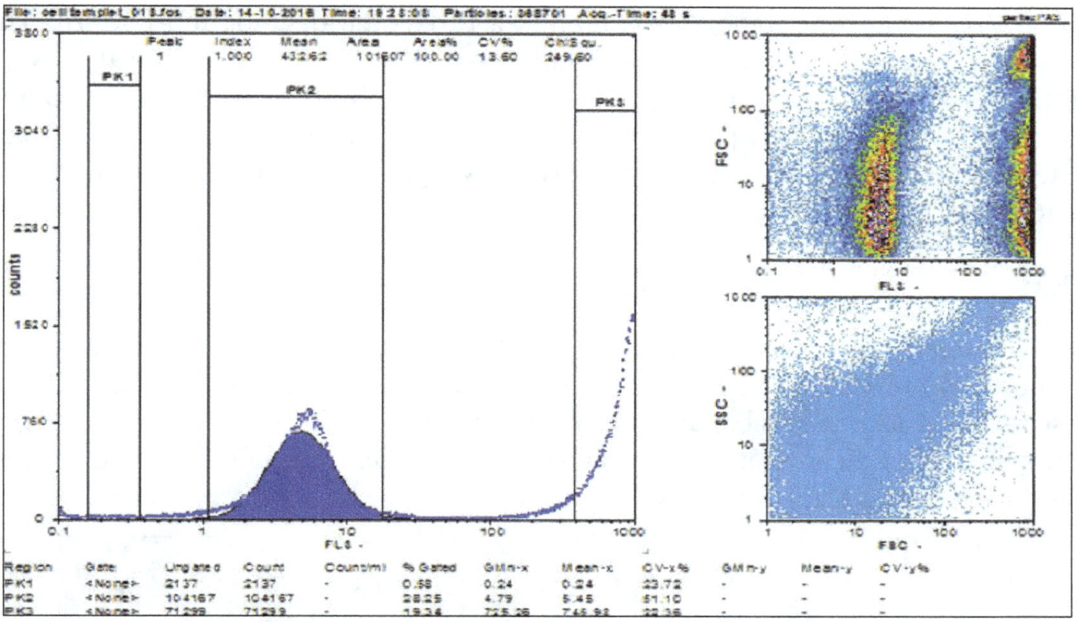

Flow cytometric analysis of compound 1, against Caco2 cell line.

Chitosan-Nanocomplex 2 Composite Nanofibers

The chitosan played an important role in the synthesis of the composite nanofiber. The amino group has been used to stabilize metal nanoparticles, so we suggested that the free NH_2 ligands of chitosan might bind to the nanoparticles and stabilized them.

FTIR

The formation of the CS-stabilized NPs was further investigated by FTIR spectroscopy. The FTIR spectra of the CS-NP nanofiber and CS nanofiber attributed a broad band at the range of 3350–3510 cm^{-1} corresponding to the amine and hydroxyl groups. For the nanocomposite, there is a change in the shape and frequencies of the bands in this region, indicating participation of the NH_2 group in the reaction. Although there is some difference in the two figures, both of them showed the basic characteristic peaks for CS at 3422 cm^{-1} (O—H stretch) and 2960–2970 cm^{-1} (C—H stretch), which indicated that the CS can adsorb on the formed nanoparticles in the preparation process.

TGA

Thermogravimetric analysis (TGA) is widely used to investigate the thermal decomposition of a polymer. The TGA thermograms of pristine CS nanofiber, CS-NP composite nanofiber, at a heating rate of 20°C/min under nitrogen atmosphere are investigated. Chitosan's nanofiber weight change between 28.27 and 100°C is associated with the loss of adsorbed and bound water. Chitosan polymer degradation started at 100°C and continued up to 333.80°C, with a 75.5% polymer weight loss.

The CS-NP composite nanofiber exhibited a three-step degradation pattern. The first step occurred between 24.68°C and 100.86°C, which is due to water evaporation. The second and the major mass loss were observed between 100°C and 261.34°C, with a 40.42% polymer weight loss. The third step occurred between 261.34 and 367.43°C, with a 19.95% polymer weight loss with total loss of 71.27% at 367.43°C.

TEM

The NP complex was encapsulated by CS, and their sizes were in the range of 15–50 nm as shown in the TEM image of the CS–NP composite nanofiber (Figure). It could be observed that a considerable amount of non-spherical particle was observed. These non-spherical particles seemed to be fabricated by coalescence of two spherical particles during the nucleation process.

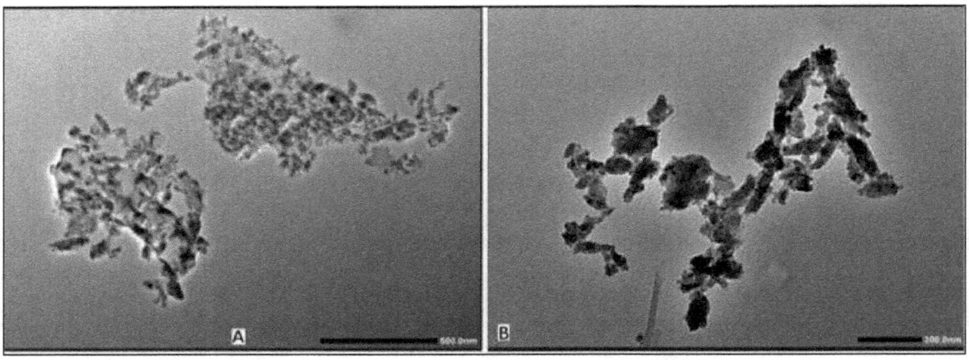

TEM micrograph of (A) chitosan nanofiber and (B) (chitosan-nanoparticle) composite nanofiber.

SEM

The morphology of electrospun (chitosan–nanoparticle) composite nanofiber is shown in figure. It is clear from these figures that no phase separation has been observed and the homogeneity of the obtained nanofibers can be easily observed and the composite of the nanoparticles has been proven. It is also clear that the diameter of chitosan-nanoparticle composite nanofiber is smaller than chitosan nanofiber.

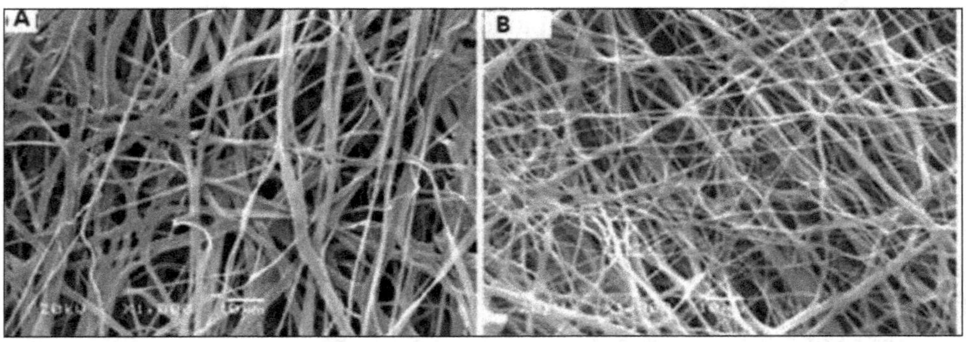

SEM micrograph 1000× of (A) CS nanofiber and (B) La-CS composite nanofiber.

Antimicrobial Activity

The ability of prepared La nanoparticle-coated nanofibers to cease the growth of microbes known for its human pathogenicity has been investigated. La nanoparticle-coated nanofibers succeeded to locally cease the growth of all tested microbes. Each disc succeeded to stop the growth of tested microbes at the boundaries between the disc and the plates' surfaces, forming clear zones with 7 mm diameter. It is worth mentioning that La nanoparticle-coated nanofibers showed the same pattern of antimicrobial activity against all tested microbes. The ability of prepared nanoparticle-nanofiber hybrid to affect the microbes locally is definitely promising for biomedical applications. These results would support the coexistence of the microbes and their antimicrobial agent at the same location and hence improve the overall microbe drug interaction process. Moreover, this kind of antimicrobial agent localization would prevent the probability of its migration to other sites inside the body that would help to avoid the expected subsidiary complications. It would be recommended that these nanoparticle-coated nanofibers are used as wound dressing for skin injuries.

Catalysis in Green Chemistry

4

- Biocatalysis
- Enzymes
- Enzyme Catalysis
- Acid Catalysts
- Heterogenous Catalysis
- Homogenous Catalysis
- Phase-transfer Catalyst
- Organocatalysis
- Photocatalysis
- Nanocatalysts
- Bifunctional Catalysts

Catalysis is one of the fundamental principles of green chemistry that produces substantial material and saves energy. A few of its aspects are biocatalysis, enzyme catalysis, heterogenous catalysis, homogenous catalysis, photocatalysis, etc. This chapter has been carefully written to provide an easy understanding of catalysis in green chemistry.

Catalyst is any substance that increases the rate of a reaction without itself being consumed. Enzymes are naturally occurring catalysts responsible for many essential biochemical reactions.

Most solid catalysts are metals or the oxides, sulfides, and halides of metallic elements and of the semimetallic elements boron, aluminum, and silicon. Gaseous and liquid catalysts are commonly used in their pure form or in combination with suitable carriers or solvents; solid catalysts are commonly dispersed in other substances known as catalyst supports.

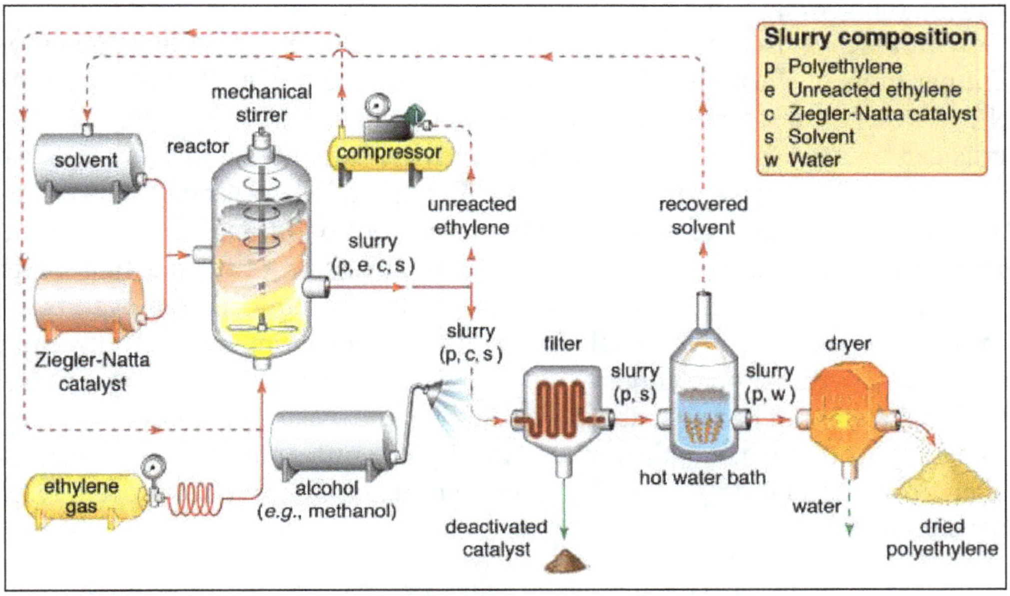

The Ziegler-Natta polymerization of ethylene.

In general, catalytic action is a chemical reaction between the catalyst and a reactant, forming chemical intermediates that are able to react more readily with each other or with another reactant, to form the desired end product. During the reaction between the chemical intermediates and the reactants, the catalyst is regenerated. The modes of reactions between the catalysts and the reactants vary widely and in solid catalysts are often complex. Typical of these reactions are acid–base reactions, oxidation–reduction reactions, formation of coordination complexes, and formation of free radicals. With solid catalysts the reaction mechanism is strongly influenced by surface properties and electronic or crystal structures. Certain solid catalysts, called polyfunctional catalysts, are capable of more than one mode of interaction with the reactants; bifunctional catalysts are used extensively for reforming reactions in the petroleum industry.

Catalyzed reactions form the basis of many industrial chemical processes. Catalyst manufacture is itself a rapidly growing industrial process.

Catalytic processes and their catalysts	
Process	Catalyst
Ammonia synthesis	Iron
Sulfuric acid manufacture	Nitrogen(II) oxide, platinum
Cracking of petroleum	Zeolites
Hydrogenation of unsaturated hydrocarbons	Nickel, platinum, or palladium
Oxidation of hydrocarbons in automobile exhausts	Copper(II) oxide, vanadium(V) oxide, platinum, palladium
Isomerization of n-butane to isobutane	Aluminum chloride, hydrogen chloride

Catalysis, one of the overarching principles of green chemistry, is capable of producing substantial material and energy savings as well as economic benefits. When comparing catalytic methods to traditional stoichiometric syntheses, catalysis offers undisputable economy in reagent use. Stoichiometric reactions would require at least one mole of reagent per mole of starting material, whereas a catalyst, subject to turnover number (TON), can carry out a transformation multiple times per mole of catalyst. Catalytic reagents can be used to improve product selectivity, by enabling diastereomeric control and site specific transformations in multi-functionalised molecules. Moreover catalysts allow reactions to proceed under milder reaction conditions. Moreover, catalytic methods can circumvent the need for pre-functionalisation of the starting materials with activating or directing groups, which would later need to be removed from the final molecule. Thus the use of catalytic rather than stoichiometric methods can dramatically enhance the atom economy of a synthesis, save time and energy, as well as decreasing the amount of raw materials consumed and waste generated. The development and application of catalytic methods can significantly improve both the economic and environmental profile when designing synthetic routes within process research and development (R&D).

There are several factors to consider when selecting a catalyst in terms of its green credentials, including:

- Catalyst efficacy eg. product yield, Turnover number (TON), space-time yield etc.
- Homogeneous or heterogeneous catalysts: While homogeneous catalysis tends to afford high activity and selectivity, heterogeneous systems benefit from ease of operational use and catalyst recovery and recycling, which has obvious advantages in terms of economy and sustainability. Catalysts can be heterogenised by the entrapping or grafting of the active molecules onto the surfaces or pores of solid supports such silica, alumina and ceria among others (Also known as immobilised catalysts).
- Chemical catalysts vs biological catalysts (enzymes): Chemical catalysts can often rely on the usage of rare and precious heavy metals to achieve the desired chemistry whilst biological catalysts can require dilute conditions and operate under a smaller range of solvents.

Biocatalysis

Biocatalysis refers to the use of living (biological) systems or their parts to speed up (catalyze) chemical reactions. In biocatalytic processes, natural catalysts, such as enzymes, perform chemical transformations on organic compounds. Both enzymes that have been more or less isolated and enzymes still residing inside living cells are employed for this task. Modern biotechnology, specifically directed evolution, has made the production of modified or non-natural enzymes possible. This has enabled the development of enzymes that can catalyze novel small molecule transformations that may be difficult or impossible using classical synthetic organic chemistry. Utilizing natural or modified enzymes to perform organic synthesis is termed chemoenzymatic synthesis; the reactions performed by the enzyme are classified as chemoenzymatic reactions.

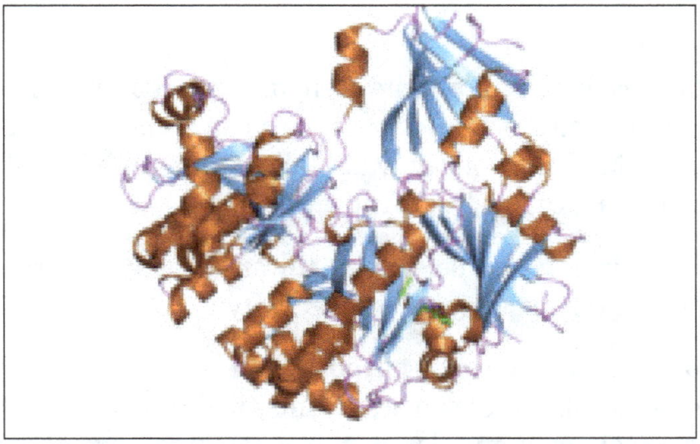

Three dimensional structure of an enzyme. Biocatalysis utilizes these biological macromolecules to catalyze small molecule transformations.

Biocatalysis underpins some of the oldest chemical transformations known to humans, for brewing predates recorded history. The oldest records of brewing are about 6000 years old and refer to the Sumerians.

The employment of enzymes and whole cells have been important for many industries for centuries. The most obvious uses have been in the food and drink businesses where the production of wine, beer, cheese etc. is dependent on the effects of the microorganisms.

More than one hundred years ago, biocatalysis was employed to do chemical transformations on non-natural man-made organic compounds, with the last 30 years seeing a substantial increase in the application of biocatalysis to produce fine chemicals, especially for the pharmaceutical industry.

Since biocatalysis deals with enzymes and microorganisms, it is historically classified separately from "homogeneous catalysis" and "heterogeneous catalysis". However, mechanistically speaking, biocatalysis is simply a special case of heterogeneous catalysis.

Advantages of Chemoenzymatic Synthesis

- Enzymes are environmentally benign, being completely degraded in the environment.

- Most enzymes typically function under mild or biological conditions, which minimizes problems of undesired side-reactions such as decomposition, isomerization, racemization and rearrangement, which often plague traditional methodology.

- Enzymes selected for chemoenzymatic synthesis can be immobilized on a solid support. These immobilized enzymes demonstrate very high stability and re-usability and can be used to conduct reactions in continuous mode in microreactors.

- Through the development of protein engineering, specifically site-directed mutagenesis and directed evolution, enzymes can be modified to enable non-natural reactivity. Modifications may also allow for a broader substrate range, enhance reaction rate or catalyst turnover.

- Enzymes exhibit extreme selectivity towards their substrates. Typically enzymes display three major types of selectivity:

    - Chemoselectivity: Since the purpose of an enzyme is to act on a single type of functional group, other sensitive functionalities, which would normally react to a certain extent under chemical catalysis, survive. As a result, biocatalytic reactions tend to be "cleaner" and laborious purification of product(s) from impurities emerging through side-reactions can largely be omitted.

    - Regioselectivity and diastereoselectivity: Due to their complex three-dimensional structure, enzymes may distinguish between functional groups which are chemically situated in different regions of the substrate molecule.

    - Enantioselectivity: Since almost all enzymes are made from L-amino acids, enzymes are chiral catalysts. As a consequence, any type of chirality present in the substrate molecule is "recognized" upon the formation of the enzyme-substrate complex. Thus a prochiral substrate may be transformed into an optically active product and both enantiomers of a racemic substrate may react at different rates.

These reasons, and especially the latter, are the major reasons why synthetic chemists have become interested in biocatalysis. This interest in turn is mainly due to the need to synthesize enantiopure compounds as chiral building blocks for Pharmaceutical drugs and agrochemicals.

Asymmetric Biocatalysis

The use of biocatalysis to obtain enantiopure compounds can be divided into two different methods:

- Kinetic resolution of a racemic mixture.

- Biocatalyzed asymmetric synthesis.

In kinetic resolution of a racemic mixture, the presence of a chiral object (the enzyme) converts one of the stereoisomers of the reactant into its product at a greater reaction rate than for the other reactant stereoisomer. The stereochemical mixture has now been transformed into a mixture of two different compounds, making them separable by normal methodology.

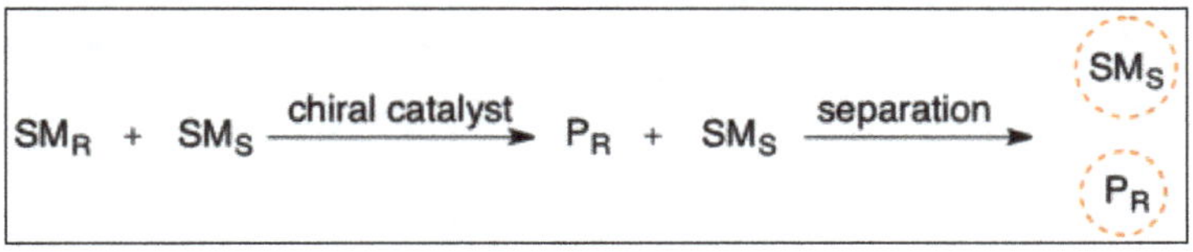

Biocatalyzed kinetic resolution is utilized extensively in the purification of racemic mixtures of synthetic amino acids. Many popular amino acid synthesis routes, such as the Strecker Synthesis, result in a mixture of R and S enantiomers. This mixture can be purified by (I) acylating the amine using an anhydride and then (II) selectively deacylating only the L enantiomer using hog kidney acylase. These enzymes are typically extremely selective for one enantiomer leading to very large differences in rate, allowing for selective deacylation. Finally the two products are now separable by classical techniques, such as chromatography.

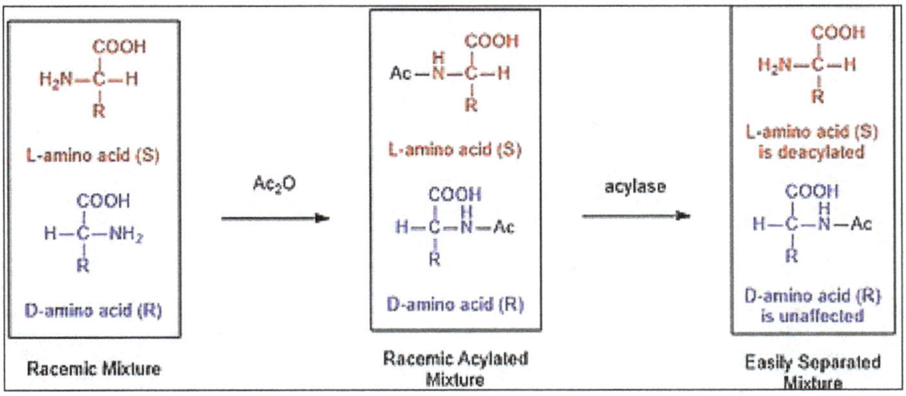

The maximum yield in such kinetic resolutions is 50%, since a yield of more than 50% means that some of wrong isomer also has reacted, giving a lower enantiomeric excess. Such reactions must therefore be terminated before equilibrium is reached. If it is possible to perform such resolutions under conditions where the two substrate- enantiomers are racemizing continuously, all substrate may in theory be converted into enantiopure product. This is called dynamic resolution.

In biocatalyzed asymmetric synthesis, a non-chiral unit becomes chiral in such a way that the different possible stereoisomers are formed in different quantities. The chirality is introduced into the substrate by influence of enzyme, which is chiral. Yeast is a biocatalyst for the enantioselective reduction of ketones.

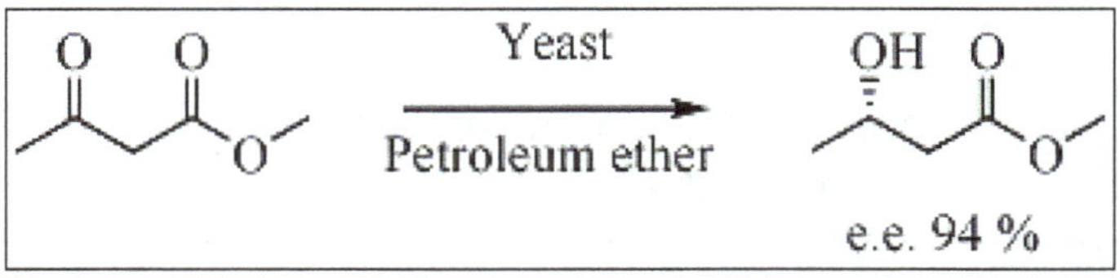

The Baeyer–Villiger oxidation is another example of a biocatalytic reaction. In one study a specially designed mutant of Candida antarctica was found to be an effective catalyst for the Michael addition of acrolein with acetylacetone at 20 °C in absence of additional solvent.

Another study demonstrates how racemic nicotine (mixture of S and R-enantiomers 1 in scheme 3) can be deracemized in a one-pot procedure involving a monoamine oxidase isolated from Aspergillus niger which is able to oxidize only the amine S-enantiomer to the imine 2 and involving an ammonia–borane reducing couple which can reduce the imine 2 back to the amine 1. In this way the S-enantiomer will continuously be consumed by the enzyme while the R-enantiomer accumulates. It is even possible to stereoinvert pure S to pure R.

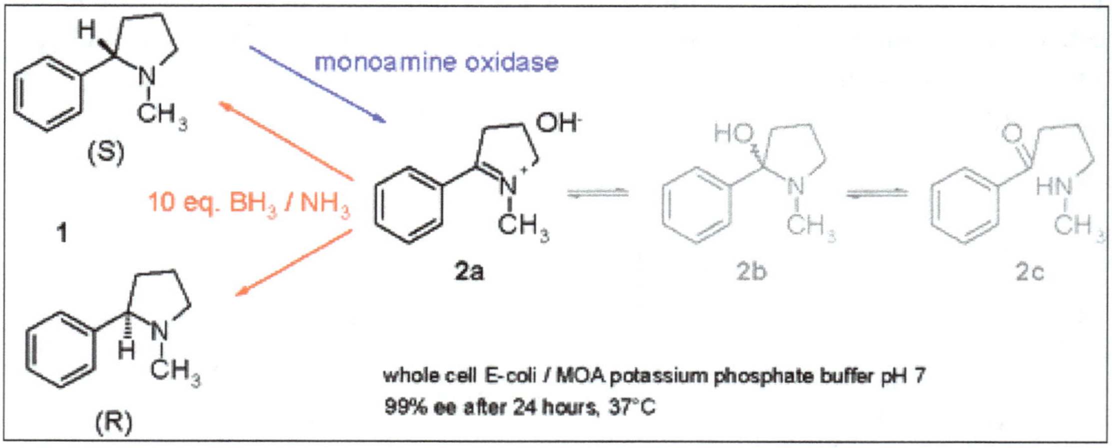

Photoredox Enabled Biocatalysis

Recently, photoredox catalysis has been applied to biocatalysis, enabling unique, previously inaccessible transformations. Photoredox chemistry relies upon light to generate free radical intermediates. These radical intermediates are achiral thus racemic mixtures of product are obtained when no external chiral environment is provided. Enzymes can provide this chiral environment within the active site and stabilize a particular conformation and favoring formation of one, enantiopure product. Photoredox enabled biocatalysis reactions fall into two categories:

- Internal coenzyme/cofactor photocatalyst.
- External photocatalyst.

Certain common hydrogen atom transfer (HAT) cofactors (NADPH and Flavin) can operate as single electron transfer (SET) reagents. Although these species are capable of HAT without irradiation, their redox potentials are enhance by nearly 2.0 V upon visible light irradiation. When paired with their respective enzymes (typically ene-reductases). This phenomenon has been utilized by chemists to develop enantioselective reduction methodologies. For example medium sized lactams can be synthesized in the chiral environment of an ene-reductase through a reductive, baldwin favored, radical cyclization terminated by enatioselective HAT from NADPH.

The second category of photoredox enabled biocatalytic reactions use an external photocatalyst (PC). Many types of PCs with a large range of redox potentials can be utilized, allowing for greater tunability of reactive compared to using a cofactor. Rose bengal, and external PC, was utilized in tandem with an oxioreductase to enantioselectively deacylate medium sized alpha-acyl-ketones.

Using an external PC has some downsides. For example, external PCs typically complicate reaction design because the PC may react with both the bound and unbound substrate. If a reaction occurs between the unbound substrate and the PC, enantioselectivity is lost and other side reactions may occur.

Enzymes

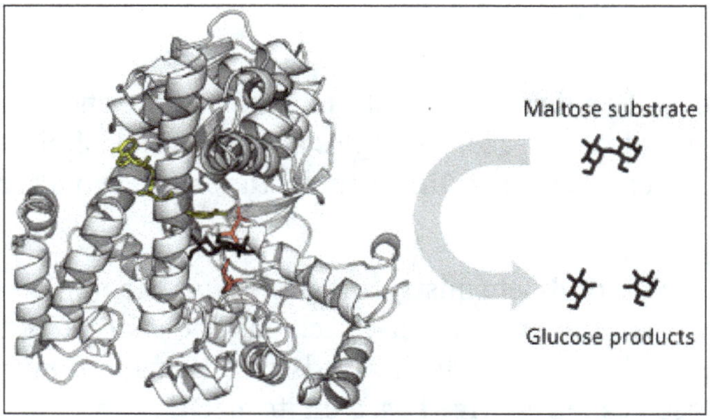

The enzyme glucosidase converts the sugar maltose into two glucose sugars.
Active site residues in red, maltose substrate in black, and NAD cofactor in yellow.

Enzymes are both proteins and biological catalysts (biocatalysts). Catalysts accelerate chemical reactions. The molecules upon which enzymes may act are called substrates, and the enzyme converts the substrates into different molecules known as products. Almost all metabolic processes in the cell need enzyme catalysis in order to occur at rates fast enough to sustain life. Metabolic pathways depend upon enzymes to catalyze individual steps. The study of enzymes is called enzymology and a new field of pseudoenzyme analysis has recently grown up, recognising that during evolution, some enzymes have lost the ability to carry out biological catalysis, which is often reflected in their amino acid sequences and unusual 'pseudocatalytic' properties.

Enzymes are known to catalyze more than 5,000 biochemical reaction types. Other biocatalysts are catalytic RNA molecules, called ribozymes. Enzymes' specificity comes from their unique three-dimensional structures.

Like all catalysts, enzymes increase the reaction rate by lowering its activation energy. Some enzymes can make their conversion of substrate to product occur many millions of times faster. An extreme example is orotidine 5'-phosphate decarboxylase, which allows a reaction that would otherwise take millions of years to occur in milliseconds. Chemically, enzymes are like any catalyst and are not consumed in chemical reactions, nor do they alter the equilibrium of a reaction. Enzymes differ from most other catalysts by being much more specific. Enzyme activity can be affected by other molecules: inhibitors are molecules that decrease enzyme activity, and activators are molecules that increase activity. Many therapeutic drugs and poisons are enzyme inhibitors. An enzyme's activity decreases markedly outside its optimal temperature and pH, and many enzymes are (permanently) denatured when exposed to excessive heat, losing their structure and catalytic properties.

Some enzymes are used commercially, for example, in the synthesis of antibiotics. Some household products use enzymes to speed up chemical reactions. Enzymes in biological washing powders break down protein, starch or fat stains on clothes, and enzymes in meat tenderizer break down proteins into smaller molecules, making the meat easier to chew.

An enzyme's name is often derived from its substrate or the chemical reaction it catalyzes, with the word ending in -ase. Examples are lactase, alcohol dehydrogenase and DNA polymerase. Different enzymes that catalyze the same chemical reaction are called isozymes.

The International Union of Biochemistry and Molecular Biology have developed a nomenclature for enzymes, the EC numbers; each enzyme is described by a sequence of four numbers preceded by "EC", which stands for "Enzyme Commission". The first number broadly classifies the enzyme based on its mechanism.

The top-level classification is:

- EC 1, Oxidoreductases: Catalyze oxidation/reduction reactions.
- EC 2, Transferases: Transfer a functional group (e.g. a methyl or phosphate group).
- EC 3, Hydrolases: Catalyze the hydrolysis of various bonds.
- EC 4, Lyases: Cleave various bonds by means other than hydrolysis and oxidation.

- EC 5, Isomerases: Catalyze isomerization changes within a single molecule.
- EC 6, Ligases: Join two molecules with covalent bonds.

These sections are subdivided by other features such as the substrate, products, and chemical mechanism. An enzyme is fully specified by four numerical designations. For example, hexokinase (EC 2.7.1.1) is a transferase (EC 2) that adds a phosphate group (EC 2.7) to a hexose sugar, a molecule containing an alcohol group (EC 2.7.1).

Structure

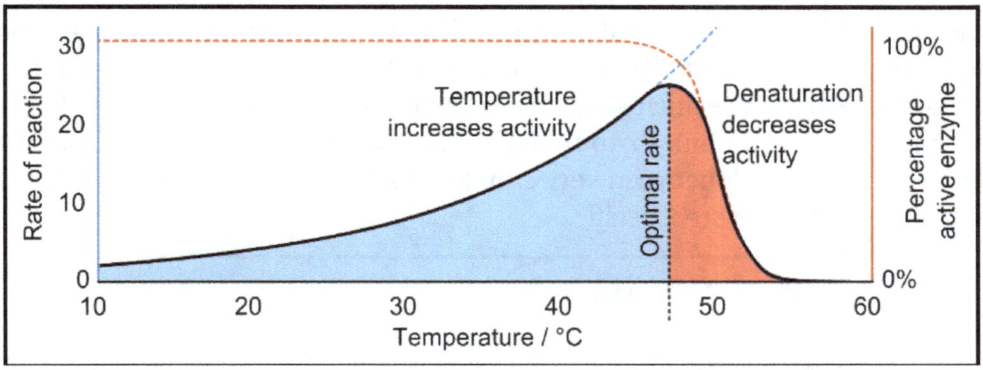

Enzyme activity initially increases with temperature (Q10 coefficient) until the enzyme's structure unfolds (denaturation), leading to an optimal rate of reaction at an intermediate temperature.

Enzymes are generally globular proteins, acting alone or in larger complexes. The sequence of the amino acids specifies the structure which in turn determines the catalytic activity of the enzyme. Although structure determines function, a novel enzymatic activity cannot yet be predicted from structure alone. Enzyme structures unfold (denature) when heated or exposed to chemical denaturants and this disruption to the structure typically causes a loss of activity. Enzyme denaturation is normally linked to temperatures above a species' normal level; as a result, enzymes from bacteria living in volcanic environments such as hot springs are prized by industrial users for their ability to function at high temperatures, allowing enzyme-catalysed reactions to be operated at a very high rate.

Enzymes are usually much larger than their substrates. Sizes range from just 62 amino acid residues, for the monomer of 4-oxalocrotonate tautomerase, to over 2,500 residues in the animal fatty acid synthase. Only a small portion of their structure (around 2–4 amino acids) is directly involved in catalysis: the catalytic site. This catalytic site is located next to one or more binding sites where residues orient the substrates. The catalytic site and binding site together comprise the enzyme's active site. The remaining majority of the enzyme structure serves to maintain the precise orientation and dynamics of the active site.

In some enzymes, no amino acids are directly involved in catalysis; instead, the enzyme contains sites to bind and orient catalytic cofactors. Enzyme structures may also contain allosteric sites where the binding of a small molecule causes a conformational change that increases or decreases activity.

A small number of RNA-based biological catalysts called ribozymes exist, which again can act alone or in complex with proteins. The most common of these is the ribosome which is a complex of protein and catalytic RNA components.

Mechanism

Substrate Binding

Enzymes must bind their substrates before they can catalyse any chemical reaction. Enzymes are usually very specific as to what substrates they bind and then the chemical reaction catalysed. Specificity is achieved by binding pockets with complementary shape, charge and hydrophilic/hydrophobic characteristics to the substrates. Enzymes can therefore distinguish between very similar substrate molecules to be chemoselective, regioselective and stereospecific.

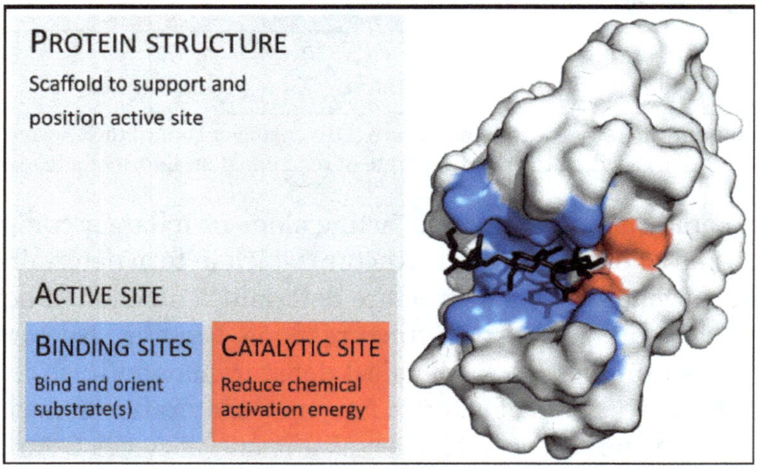

Organisation of enzyme structure and lysozyme example. Binding sites in blue, catalytic site in red and peptidoglycan substrate in black.

Some of the enzymes showing the highest specificity and accuracy are involved in the copying and expression of the genome. Some of these enzymes have "proof-reading" mechanisms. Here, an enzyme such as DNA polymerase catalyzes a reaction in a first step and then checks that the product is correct in a second step. This two-step process results in average error rates of less than 1 error in 100 million reactions in high-fidelity mammalian polymerases. Similar proofreading mechanisms are also found in RNA polymerase, aminoacyl tRNA synthetases and ribosomes.

Conversely, some enzymes display enzyme promiscuity, having broad specificity and acting on a range of different physiologically relevant substrates. Many enzymes possess small side activities which arose fortuitously (i.e. neutrally), which may be the starting point for the evolutionary selection of a new function.

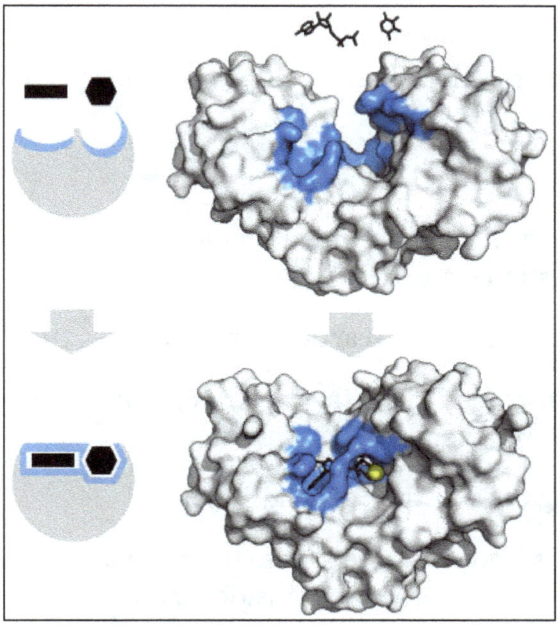

Enzyme changes shape by induced fit upon substrate binding to form enzyme-substrate complex. Hexokinase has a large induced fit motion that closes over the substrates adenosine triphosphate and xylose. Binding sites in blue, substrates in black and Mg^{2+} cofactor in yellow.

"Lock and Key" Model

To explain the observed specificity of enzymes, in 1894 Emil Fischer proposed that both the enzyme and the substrate possess specific complementary geometric shapes that fit exactly into one another. This is often referred to as "the lock and key" model. This early model explains enzyme specificity, but fails to explain the stabilization of the transition state that enzymes achieve.

Induced Fit Model

In 1958, Daniel Koshland suggested a modification to the lock and key model: since enzymes are rather flexible structures, the active site is continuously reshaped by interactions with the substrate as the substrate interacts with the enzyme. As a result, the substrate does not simply bind to a rigid active site; the amino acid side-chains that make up the active site are molded into the precise positions that enable the enzyme to perform its catalytic function. In some cases, such as glycosidases, the substrate molecule also changes shape slightly as it enters the active site. The active site continues to change until the substrate is completely bound, at which point the final shape and charge distribution is determined. Induced fit may enhance the fidelity of molecular recognition in the presence of competition and noise via the conformational proofreading mechanism.

Catalysis

Enzymes can accelerate reactions in several ways, all of which lower the activation energy ($\Delta G^{\ddagger}$, Gibbs free energy).

- By stabilizing the transition state:
    - Creating an environment with a charge distribution complementary to that of the transition state to lower its energy.
- By providing an alternative reaction pathway:
    - Temporarily reacting with the substrate, forming a covalent intermediate to provide a lower energy transition state.
- By destabilising the substrate ground state:
    - Distorting bound substrate(s) into their transition state form to reduce the energy required to reach the transition state.
    - By orienting the substrates into a productive arrangement to reduce the reaction entropy change (the contribution of this mechanism to catalysis is relatively small).

Enzymes may use several of these mechanisms simultaneously. For example, proteases such as trypsin perform covalent catalysis using a catalytic triad, stabilise charge build-up on the transition states using an oxyanion hole, complete hydrolysis using an oriented water substrate.

Dynamics

Enzymes are not rigid, static structures; instead they have complex internal dynamic motions – that is, movements of parts of the enzyme's structure such as individual amino acid residues, groups of residues forming a protein loop or unit of secondary structure, or even an entire protein domain. These motions give rise to a conformational ensemble of slightly different structures that interconvert with one another at equilibrium. Different states within this ensemble may be associated with different aspects of an enzyme's function. For example, different conformations of the enzyme dihydrofolate reductase are associated with the substrate binding, catalysis, cofactor release, and product release steps of the catalytic cycle, consistent with catalytic resonance theory.

Substrate Presentation

Substrate presentation is a process where the enzyme is sequestered away from its substrate. Enzymes can be sequestered to the plasma membrane away from a substrate in the nucleus or cytosol. Or within the membrane, an enzyme can be sequestered into lipid rafts away from its substrate in the disordered region. When the enzyme is releases it mixes with its substrate. Alternatively, the enzyme can be sequestered near its substrate to activate the enzyme. For example, the enzyme can be soluble and upon activation bind to a lipid in the plasma membrane and then act upon molecules in the plasma membrane.

Allosteric Modulation

Allosteric sites are pockets on the enzyme, distinct from the active site, that bind to molecules in the cellular environment. These molecules then cause a change in the conformation or dynamics of the enzyme that is transduced to the active site and thus affects the reaction rate of the enzyme. In this way, allosteric interactions can either inhibit or activate enzymes. Allosteric interactions with metabolites upstream or downstream in an enzyme's metabolic pathway cause feedback regulation, altering the activity of the enzyme according to the flux through the rest of the pathway.

Cofactors

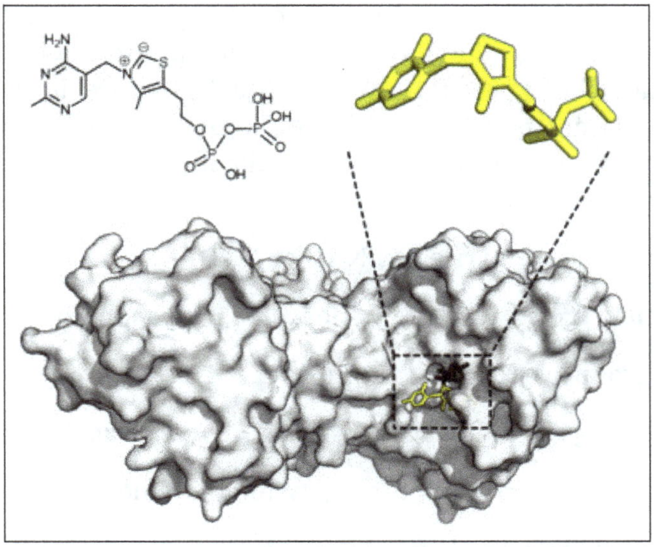

Chemical structure for thiamine pyrophosphate and protein structure of transketolase. Thiamine pyrophosphate cofactor in yellow and xylulose 5-phosphate substrate in black.

Some enzymes do not need additional components to show full activity. Others require non-protein molecules called cofactors to be bound for activity. Cofactors can be either inorganic (e.g., metal ions and iron-sulfur clusters) or organic compounds (e.g., flavin and heme). These cofactors serve many purposes; for instance, metal ions can help in stabilizing nucleophilic species within the active site. Organic cofactors can be either coenzymes, which are released from the enzyme's active site during the reaction, or prosthetic groups, which are tightly bound to an enzyme. Organic prosthetic groups can be covalently bound (e.g., biotin in enzymes such as pyruvate carboxylase).

An example of an enzyme that contains a cofactor is carbonic anhydrase, which uses a zinc cofactor bound as part of its active site. These tightly bound ions or molecules are usually found in the active site and are involved in catalysis. For example, flavin and heme cofactors are often involved in redox reactions.

Enzymes that require a cofactor but do not have one bound are called apoenzymes or apoproteins. An enzyme together with the cofactor(s) required for activity is called a holoenzyme (or haloenzyme). The term holoenzyme can also be applied to enzymes that contain multiple protein subunits, such as the DNA polymerases; here the holoenzyme is the complete complex containing all the subunits needed for activity.

Coenzymes

Coenzymes are small organic molecules that can be loosely or tightly bound to an enzyme. Coenzymes transport chemical groups from one enzyme to another. Examples include NADH, NADPH and adenosine triphosphate (ATP). Some coenzymes, such as flavin mononucleotide (FMN), flavin adenine dinucleotide (FAD), thiamine pyrophosphate (TPP), and tetrahydrofolate (THF), are derived from vitamins. These coenzymes cannot be synthesized by the body de novo and closely related compounds (vitamins) must be acquired from the diet. The chemical groups carried include:

- The hydride ion (H^-), carried by NAD or $NADP^+$.
- The phosphate group, carried by adenosine triphosphate.
- The acetyl group, carried by coenzyme A.
- Formyl, methenyl or methyl groups, carried by folic acid.
- The methyl group, carried by S-adenosylmethionine.

Since coenzymes are chemically changed as a consequence of enzyme action, it is useful to consider coenzymes to be a special class of substrates, or second substrates, which are common to many different enzymes. For example, about 1000 enzymes are known to use the coenzyme NADH.

Coenzymes are usually continuously regenerated and their concentrations maintained at a steady level inside the cell. For example, NADPH is regenerated through the pentose phosphate pathway and S-adenosylmethionine by methionine adenosyltransferase. This continuous regeneration means that small amounts of coenzymes can be used very intensively. For example, the human body turns over its own weight in ATP each day.

Thermodynamics

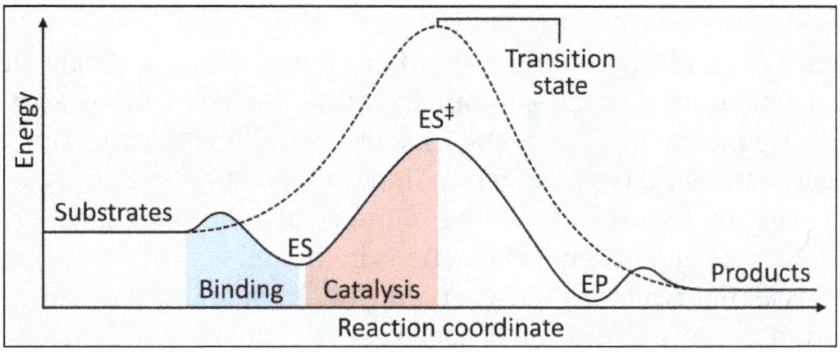

The energies of the stages of a chemical reaction. Uncatalysed (dashed line), substrates need a lot of activation energy to reach a transition state, which then decays into lower-energy products. When enzyme catalysed (solid line), the enzyme binds the substrates (ES), then stabilizes the transition state ($ES^{\ddagger}$) to reduce the activation energy required to produce products (EP) which are finally released.

As with all catalysts, enzymes do not alter the position of the chemical equilibrium of the reaction. In the presence of an enzyme, the reaction runs in the same direction as it would without the enzyme, just more quickly. For example, carbonic anhydrase catalyzes its reaction in either direction depending on the concentration of its reactants:

$$CO_2 + H_2O \xrightarrow{\text{Carbonic anhydrase}} H_2CO_3 \text{ (in tissues; high } CO_2 \text{ concentration)}$$

$$CO_2 + H_2O \xleftarrow{\text{Carbonic anhydrase}} H_2CO_3 \text{ (in lungs; low } CO_2 \text{ concentration))}$$

The rate of a reaction is dependent on the activation energy needed to form the transition state which then decays into products. Enzymes increase reaction rates by lowering the energy of the transition state. First, binding forms a low energy enzyme-substrate complex (ES). Secondly the enzyme stabilises the transition state such that it requires less energy to achieve compared to the uncatalyzed reaction ($ES^{\ddagger}$). Finally the enzyme-product complex (EP) dissociates to release the products.

Enzymes can couple two or more reactions, so that a thermodynamically favorable reaction can be used to "drive" a thermodynamically unfavourable one so that the combined energy of the products is lower than the substrates. For example, the hydrolysis of ATP is often used to drive other chemical reactions.

Enzyme Catalysis

Enzyme catalysis is the increase in the rate of a process by a biological molecule, an "enzyme". Most enzymes are proteins, and most such processes are chemical reactions. Within the enzyme, generally catalysis occurs at a localized site, called the active site.

Most enzymes are made predominantly of proteins, either a single protein chain or many such chains in a multi-subunit complex. Enzymes often also incorporate non-protein components, such as metal ions or specialized organic molecules known as cofactor (e.g. adenosine triphosphate). Many cofactors are vitamins, and their role as vitamins is directly linked to their use in the catalysis of biological process within metabolism. Catalysis of biochemical reactions in the cell is vital since many but not all metabolically essential reactions have very low rates when uncatalysed. One driver of protein evolution is the optimization of such catalytic activities, although only the most crucial enzymes operate near catalytic efficiency limits, and many enzymes are far from optimal. Important factors in enzyme catalysis include general acid and base catalysis, orbital steering, entropic restriction, orientation effects (i.e. lock and key catalysis), as well as motional effects involving protein dynamics.

Mechanisms of enzyme catalysis vary, but are all similar in principle to other types of chemical catalysis in that the crucial factor is a reduction of energy barrier(s) separating the reactants from the products. The reduction of activation energy (E_a) increases the fraction of reactant molecules that can overcome this barrier and form the product. An important principle is that since they only reduce energy barriers between products and reactants, enzymes always catalyze reactions in both directions, and cannot drive a reaction forward or affect the equilibrium position - only the speed with which is it achieved. As with other catalysts, the enzyme is not consumed or changed by the reaction (as a substrate is) but is recycled such that a single enzyme performs many rounds of catalysis.

Induced Fit

The favored model for the enzyme-substrate interaction is the induced fit model. This model proposes that the initial interaction between enzyme and substrate is relatively weak, but that these weak interactions rapidly induce conformational changes in the enzyme that strengthen binding.

The advantages of the induced fit mechanism arise due to the stabilizing effect of strong enzyme binding. There are two different mechanisms of substrate binding: uniform binding, which has strong substrate binding, and differential binding, which has strong transition state binding. The stabilizing effect of uniform binding increases both substrate and transition state binding affinity, while differential binding increases only transition state binding affinity. Both are used by enzymes and have been evolutionarily chosen to minimize the activation energy of the reaction. Enzymes that are saturated, that is, have a high affinity substrate binding, require differential binding to reduce the energy of activation, whereas small substrate unbound enzymes may use either differential or uniform binding.

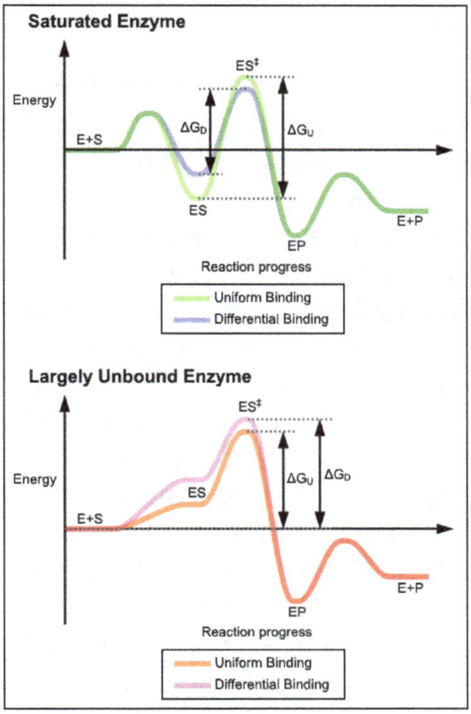

The different mechanisms of substrate binding.

These effects have led to most proteins using the differential binding mechanism to reduce the energy of activation, so most substrates have high affinity for the enzyme while in the transition state. Differential binding is carried out by the induced fit mechanism - the substrate first binds weakly, then the enzyme changes conformation increasing the affinity to the transition state and stabilizing it, so reducing the activation energy to reach it.

It is important to clarify, however, that the induced fit concept cannot be used to rationalize catalysis. That is, the chemical catalysis is defined as the reduction of $E_a^‡$ (when the system is already in the $ES^‡$) relative to $E_a^‡$ in the uncatalyzed reaction in water (without the enzyme). The induced fit only suggests that the barrier is lower in the closed form of the enzyme but does not tell us what the reason for the barrier reduction is. Induced fit may be beneficial to the fidelity of molecular recognition in the presence of competition and noise via the conformational proofreading mechanism.

Mechanisms of an Alternative Reaction Route

These conformational changes also bring catalytic residues in the active site close to the chemical bonds in the substrate that will be altered in the reaction. After binding takes place, one or more mechanisms of catalysis lowers the energy of the reaction's transition state, by providing an alternative chemical pathway for the reaction. There are six possible mechanisms of "over the barrier" catalysis as well as a "through the barrier" mechanism.

Proximity and Orientation

Enzyme-substrate interactions align the reactive chemical groups and hold them close together in an optimal geometry, which increases the rate of the reaction. This reduces the entropy of the reactants and thus makes addition or transfer reactions less unfavorable, since a reduction in the overall entropy when two reactants become a single product. However this is a general effect and is seen in non-addition or transfer reactions where it occurs due to an increase in the "effective concentration" of the reagents. This is understood when considering how increases in concentration leads to increases in reaction rate. Essentially when the reactants are more concentrated, they collide more often and so react more often. In enzyme catalysis, the binding of the reagents to the enzyme restricts the conformational space of the reactants, holding them in the 'proper orientation' and close to each other, so that the collide more frequently, and with the correct geometry, to facilitate the desired reaction. The "effective concentration" is the concentration the reactant would have to be, free in solution, to experiences the same collisional frequency. Often such theoretical effective concentrations are unphysical and impossible to realize in reality - which is a testament to the great catalytic power of many enzymes, with massive rate increases over the uncatalyzed state.

For example:

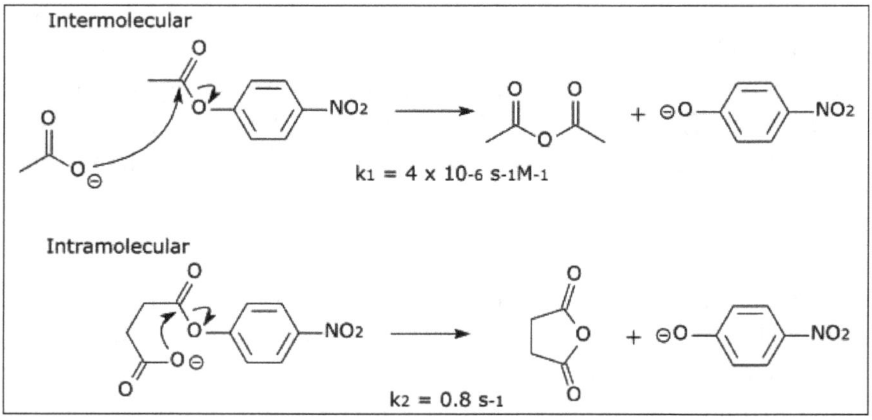

The effective concentration of acetate in the intramolecular reaction can be estimated as $k_2/k_1 = 2 \times 10^5$ Molar.

However, the situation might be more complex, since modern computational studies have established that traditional examples of proximity effects cannot be related directly to enzyme entropic effects. Also, the original entropic proposal has been found to largely overestimate the contribution of orientation entropy to catalysis.

Proton Donors or Acceptors

Proton donors and acceptors, i.e. acids and base may donate and accept protons in order to stabilize developing charges in the transition state. This is related to the overall principle of catalysis, that of reducing energy barriers, since in general transition stated are high energy states, and by stabilizing them this high energy is reduced, lowering the barrier. A key feature of enzyme catalysis over many non-biological catalysis, is that both acid and base catalysis can be combined in the same reaction. In many abiotic systems, acids (large [H+]) or bases (large concentration H+ sinks, or species with electron pairs) can increase the rate of the reaction; but of course the environment can only have one overall pH (measure of acidity or basicity (alkalinity)). However since enzymes are large molecules, they can position both acid groups and basic groups in their active site to interact with their substrates, and employ both modes independent of the bulk pH.

Often general acid or base catalysis is employed to activate nucleophile and/or electrophile groups, or to stabilizing leaving groups. Many amino acids with acidic or basic groups are this employed in the active site, such as the glutamic and aspartic acid, histidine, cystine, tyrosine, lysine and arginine, as well as serine and threonine. In addition, the peptide backbone, with carbonyl and amide N groups is often employed. Cystine and Histidine are very commonly involved, since they both have a pKa close to neutral pH and can therefore both accept and donate protons.

Many reaction mechanisms involving acid/base catalysis assume a substantially altered pKa. This alteration of pKa is possible through the local environment of the residue.

Conditions	Acids	Bases
Hydrophobic environment	Increase pKa	Decrease pKa
Adjacent residues of like charge	Increase pKa	Decrease pKa
Salt bridge (and hydrogen bond) formation	Decrease pKa	Increase pKa

pKa can also be influenced significantly by the surrounding environment, to the extent that residues which are basic in solution may act as proton donors, and vice versa.

For example:

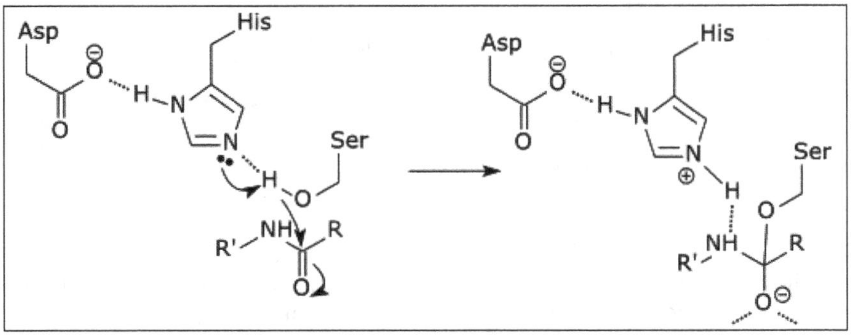

The initial step of the serine protease catalytic mechanism involves the histidine of the active site accepting a proton from the serine residue. This prepares the serine as a nucleophile to attack the amide bond of the substrate. This mechanism includes donation of a proton from serine (a base, pKa 14) to histidine (an acid, pKa 6), made possible due to the local environment of the bases.

It is important to clarify that the modification of the pKa's is a pure part of the electrostatic mechanism. Furthermore, the catalytic effect of the above example is mainly associated with the reduction of the pKa of the oxyanion and the increase in the pKa of the histidine, while the proton transfer from the serine to the histidine is not catalyzed significantly, since it is not the rate determining barrier.

Electrostatic Catalysis

Stabilization of charged transition states can also be by residues in the active site forming ionic bonds (or partial ionic charge interactions) with the intermediate. These bonds can either come from acidic or basic side chains found on amino acids such as lysine, arginine, aspartic acid or glutamic acid or come from metal cofactors such as zinc. Metal ions are particularly effective and can reduce the pKa of water enough to make it an effective nucleophile.

Systematic computer simulation studies established that electrostatic effects give, by far, the largest contribution to catalysis. In particular, it has been found that enzyme provides an environment which is more polar than water, and that the ionic transition states are stabilized by fixed dipoles. This is very different from transition state stabilization in water, where the water molecules must pay with "reorganization energy". In order to stabilize ionic and charged states. Thus, the catalysis is associated with the fact that the enzyme polar groups are preorganized. The magnitude of the electrostatic field exerted by an enzyme's active site has been shown to be highly correlated with the enzyme's catalytic rate enhancement.

For example:

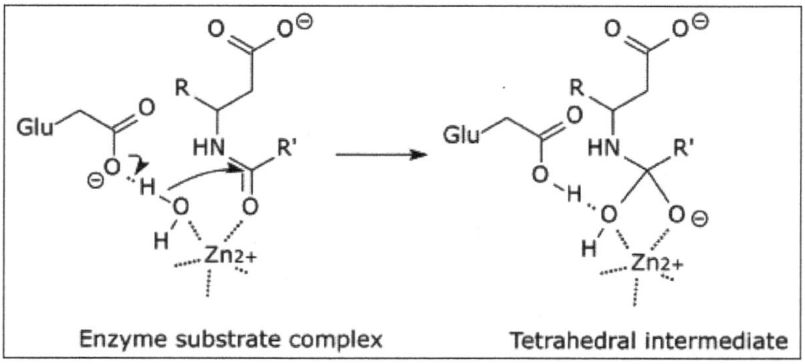

The tetrahedral intermediate is stabilised by a partial ionic bond between the Zn^{2+} ion and the negative charge on the oxygen.

Binding of substrate usually excludes water from the active site, thereby lowering the local dielectric constant to that of an organic solvent. This strengthens the electrostatic interactions between the charged/polar substrates and the active sites. In addition, studies have shown that the charge distributions about the active sites are arranged so as to stabilize the transition states of the catalyzed reactions. In several enzymes, these charge distributions apparently serve to guide polar substrates toward their binding sites so that the rates of these enzymatic reactions are greater than their apparent diffusion-controlled limits.

Covalent Catalysis

Covalent catalysis involves the substrate forming a transient covalent bond with residues in the enzyme active site or with a cofactor. This adds an additional covalent intermediate to the reaction, and helps to reduce the energy of later transition states of the reaction. The covalent bond must, at a later stage in the reaction, be broken to regenerate the enzyme. This mechanism is utilised by the catalytic triad of enzymes such as proteases like chymotrypsin and trypsin, where an acyl-enzyme intermediate is formed. An alternative mechanism is schiff base formation using the free amine from a lysine residue, as seen in the enzyme aldolase during glycolysis.

Some enzymes utilize non-amino acid cofactors such as pyridoxal phosphate (PLP) or thiamine pyrophosphate (TPP) to form covalent intermediates with reactant molecules. Such covalent intermediates function to reduce the energy of later transition states, similar to how covalent intermediates formed with active site amino acid residues allow stabilization, but the capabilities of cofactors allow enzymes to carryout reactions that amino acid side residues alone could not. Enzymes utilizing such cofactors include the PLP-dependent enzyme aspartate transaminase and the TPP-dependent enzyme pyruvate dehydrogenase.

Rather than lowering the activation energy for a reaction pathway, covalent catalysis provides an alternative pathway for the reaction (via to the covalent intermediate) and so is distinct from true catalysis. For example, the energetics of the covalent bond to the serine molecule in chymotrypsin should be compared to the well-understood covalent bond to the nucleophile in the uncatalyzed solution reaction. A true proposal of a covalent catalysis (where the barrier is lower than the corresponding barrier in solution) would require, for example, a partial covalent bond to the transition state by an enzyme group (e.g., a very strong hydrogen bond), and such effects do not contribute significantly to catalysis.

Metal Ion Catalysis

A metal ion in the active site participates in catalysis by coordinating charge stabilization and shielding. Because of a metal's positive charge, only negative charges can be stabilized through metal ions. However, metal ions are advantageous in biological catalysis because they are not affected by changes in pH. Metal ions can also act to ionize water by acting as a Lewis acid. Metal ions may also be agents of oxidation and reduction.

Bond Strain

This is the principal effect of induced fit binding, where the affinity of the enzyme to the transition state is greater than to the substrate itself. This induces structural rearrangements which strain substrate bonds into a position closer to the conformation of the transition state, so lowering the energy difference between the substrate and transition state and helping catalyze the reaction.

However, the strain effect is, in fact, a ground state destabilization effect, rather than transition state stabilization effect. Furthermore, enzymes are very flexible and they cannot apply large strain effect.

In addition to bond strain in the substrate, bond strain may also be induced within the enzyme itself to activate residues in the active site.

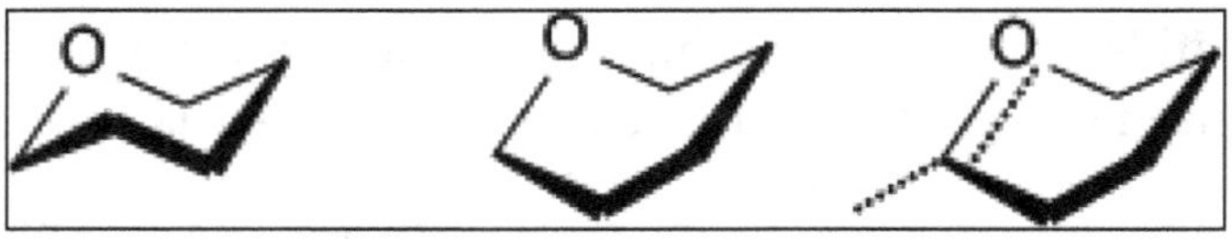

The substrate, on binding, is distorted from the half chair conformation of the hexose ring (because of the steric hindrance with amino acids of the protein forcing the equatorial c6 to be in the axial position) into the chair conformation.

Quantum Tunneling

These traditional "over the barrier" mechanisms have been challenged in some cases by models and observations of "through the barrier" mechanisms (quantum tunneling). Some enzymes operate with kinetics which are faster than what would be predicted by the classical $\Delta G^{\ddagger}$. In "through the barrier" models, a proton or an electron can tunnel through activation barriers. Quantum tunneling for protons has been observed in tryptamine oxidation by aromatic amine dehydrogenase.

Quantum tunneling does not appear to provide a major catalytic advantage, since the tunneling contributions are similar in the catalyzed and the uncatalyzed reactions in solution. However, the tunneling contribution (typically enhancing rate constants by a factor of ~1000 compared to the rate of reaction for the classical 'over the barrier' route) is likely crucial to the viability of biological organisms. This emphasizes the general importance of tunneling reactions in biology.

Active Enzyme

The binding energy of the enzyme-substrate complex cannot be considered as an external energy which is necessary for the substrate activation. The enzyme of high energy content may firstly transfer some specific energetic group X_1 from catalytic site of the enzyme to the final place of the first bound reactant, then another group X_2 from the second bound reactant (or from the second group of the single reactant) must be transferred to active site to finish substrate conversion to product and enzyme regeneration.

We can present the whole enzymatic reaction as a two coupling reactions:

$$S_1 + EX_1 \rightarrow S_1EX_1 \rightarrow P_1 + EP_2$$

$$S_2 + EP_2 \rightarrow S_2EP_2 \rightarrow P_2 + EX_2$$

It may be seen from reaction ($S_1 + EX_1 \rightarrow S_1EX_1 \rightarrow P_1 + EP_2$) that the group X_1 of the active enzyme appears in the product due to possibility of the exchange reaction inside enzyme to avoid both electrostatic inhibition and repulsion of atoms. So we represent the active enzyme as a powerful reactant of the enzymatic reaction. The reaction ($S_2 + EP_2 \rightarrow S_2EP_2 \rightarrow P_2 + EX_2$) shows incomplete conversion of the substrate because its group X_2 remains inside enzyme. This approach as idea had formerly proposed relying on the hypothetical extremely high enzymatic conversions (catalytically perfect enzyme).

The crucial point for the verification of the present approach is that the catalyst must be a complex of the enzyme with the transfer group of the reaction. This chemical aspect is supported by the well-studied mechanisms of the several enzymatic reactions. Consider the reaction of peptide bond hydrolysis catalyzed by a pure protein α-chymotrypsin (an enzyme acting without a cofactor), which is a well-studied member of the serine proteases family.

We present the experimental results for this reaction as two chemical steps:

$$S_1 + EH \rightarrow P_1 + EP_2$$

$$EP_2 + H-O-H \rightarrow EH + P_2$$

where S_1 is a polypeptide, P_1 and P_2 are products. The first chemical step ($S_1 + EH \rightarrow P_1 + EP_2$) includes the formation of a covalent acyl-enzyme intermediate. The second step ($EP_2 + H-O-H \rightarrow EH + P_2$) is the deacylation step. It is important to note that the group H+, initially found on the enzyme, but not in water, appears in the product before the step of hydrolysis, therefore it may be considered as an additional group of the enzymatic reaction.

Thus, the reaction ($S_1 + EH \rightarrow P_1 + EP_2$) shows that the enzyme acts as a powerful reactant of the reaction. the H transport from the enzyme promotes the first reactant conversion, breakdown of the first initial chemical bond (between groups P_1 and P_2). The step of hydrolysis leads to a breakdown of the second chemical bond and regeneration of the enzyme.

The proposed chemical mechanism does not depend on the concentration of the substrates or products in the medium. However, a shift in their concentration mainly causes free energy changes in the first and final steps of the reactions ($S_1 + EX_1 \rightarrow S_1EX_1 \rightarrow P_1 + EP_2$) and ($S_2 + EP_2 \rightarrow S_2EP_2 \rightarrow P_2 + EX_2$) due to the changes in the free energy content of every molecule, whether S or P, in water solution. This approach is in accordance with the following mechanism of muscle contraction. The final step of ATP hydrolysis in skeletal muscle is the product release caused by the association of myosin heads with actin. The closing of the actin-binding cleft during the association reaction is structurally coupled with the opening of the nucleotide-binding pocket on the myosin active site.

Notably, the final steps of ATP hydrolysis include the fast release of phosphate and the slow release of ADP. The release of a phosphate anion from bound ADP anion into water solution may be considered as an exergonic reaction because the phosphate anion has low molecular mass.

Thus, we arrive at the conclusion that the primary release of the inorganic phosphate $H_2PO_4^-$ leads to transformation of a significant part of the free energy of ATP hydrolysis into the kinetic energy of the solvated phosphate, producing active streaming. This assumption of a local mechano-chemical transduction is in accord with Tirosh's mechanism of muscle contraction, where the muscle force derives from an integrated action of active streaming created by ATP hydrolysis.

Examples of Catalytic Mechanisms

In reality, most enzyme mechanisms involve a combination of several different types of catalysis:

- Triose Phosphate Isomerase: Triose phosphate isomerase (EC 5.3.1.1) catalyses the reversible interconvertion of the two triose phosphates isomers dihydroxyacetone phosphate and D-glyceraldehyde 3-phosphate.
- Trypsin: Trypsin (EC 3.4.21.4) is a serine protease that cleaves protein substrates after lysine or arginine residues using a catalytic triad to perform covalent catalysis, and an oxyanion hole to stabilise charge-buildup on the transition states.
- Aldolase: Aldolase (EC 4.1.2.13) catalyses the breakdown of fructose 1,6-bisphosphate (F-1,6-BP) into glyceraldehyde 3-phosphate and dihydroxyacetone phosphate (DHAP).
- Enzyme Diffusivity: The advent of single-molecule studies led to the observation that the movement of untethered enzymes increases with increasing substrate concentration and increasing reaction enthalpy. Subsequent observations suggest that this increase in diffusivity is driven by transient displacement of the enzyme's center of mass, resulting in a "recoil effect that propels the enzyme".

- Reaction Similarity: Similarity between enzymatic reactions (EC) can be calculated by using bond changes, reaction centres or substructure metrics (EC-BLAST).

Acid Catalysts

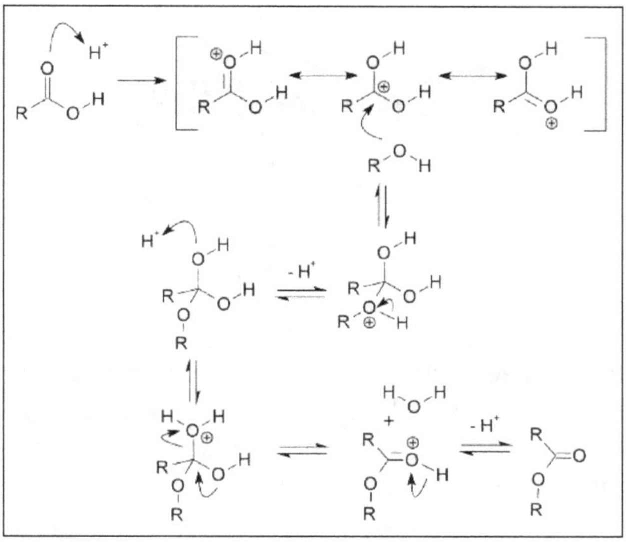

In acid-catalyzed Fischer esterification, the proton binds to oxygens and functions as a Lewis acid to activate the ester carbonyl (top row) as an electrophile, and converts the hydroxyl into the good leaving group water (bottom left). Both lower the kinetic barrier and speed up the attainment of chemical equilibrium.

In acid catalysis and base catalysis, a chemical reaction is catalyzed by an acid or a base. By Brønsted–Lowry acid–base theory, the acid is the proton (hydrogen ion, H^+) donor and the base is the proton acceptor. Typical reactions catalyzed by proton transfer are esterfications and aldol reactions. In these reactions, the conjugate acid of the carbonyl group is a better electrophile than the neutral carbonyl group itself. Depending on the chemical species that act as the acid or base, catalytic mechanisms can be classified as either specific catalysis and general catalysis. Many enzymes operate by specific catalysis.

Applications and Examples

Brønsted Acids

Acid catalysis is mainly used for organic chemical reactions. Many acids can function as sources for the protons. Acid used for acid catalysis include hydrofluoric acid (in the alkylation process), phosphoric acid, toluenesulfonic acid, polystyrene sulfonate, heteropoly acids, zeolites.

Strong acids catalyze the hydrolysis and transesterification of esters, e.g. for processing fats into biodiesel. In terms of mechanism, the carbonyl oxygen is susceptible to protonation, which enhances the electrophilicity at the carbonyl carbon.

Solid Acid Catalysts

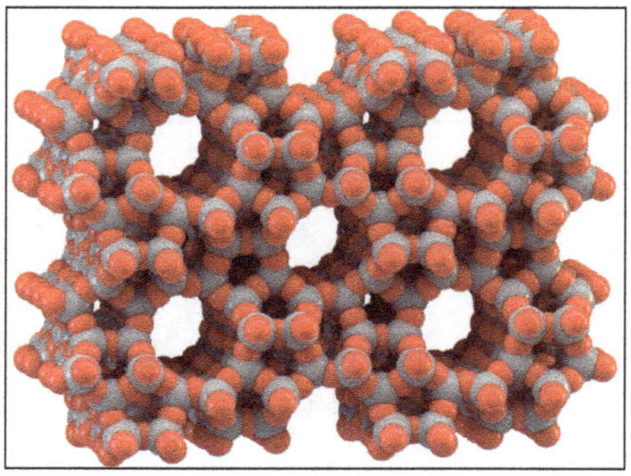

Zeolite, ZSM-5 is widely used as a solid acid catalyst.

In industrial scale chemistry, many processes are catalysed by "solid acids". As heterogeneous catalysts, solid acids do not dissolve in the reaction medium. Well known examples include these oxides, which function as Lewis acids: silico-aluminates (zeolites, alumina, silico-alumino-phosphate), sulfated zirconia, and many transition metal oxides (titania, zirconia, niobia, and more). Such acids are used in cracking. Many solid Brønsted acids are also employed industrially, including sulfonated polystyrene, solid phosphoric acid, niobic acid, and heteropolyoxometallates.

A particularly large scale application is alkylation, e.g., the combination of benzene and ethylene to give ethylbenzene. Another major application is the rearrangement of cyclohexanone oxime to caprolactam. Many alkylamines are prepared by amination of alcohols, catalyzed by solid acids. In this role, the acid converts, OH^-, a poor leaving group, into a good one. Thus acids are used to convert alcohols into other classes of compounds, such as thiols and amines.

Mechanism

Two kinds of acid catalysis are recognized, specific acid catalysis and general acid catalysis.

Specific Catalysis

In specific acid catalysis, protonated solvent is the catalyst. The reaction rate is proportional to the concentration of the protonated solvent molecules SH^+. The acid catalyst itself (AH) only contributes to the rate acceleration by shifting the chemical equilibrium between solvent S and AH in favor of the SH^+ species. This kind of catalysis is common for strong acids in polar solvents, such as water:

$$S + AH \rightarrow SH^+ + A^-$$

For example, in an aqueous buffer solution the reaction rate for reactants R depends on the pH of the system but not on the concentrations of different acids.

$$\text{rate} = -\frac{d[R^1]}{dt} = k[SH+][R^1][R^2]$$

This type of chemical kinetics is observed when reactant R^1 is in a fast equilibrium with its conjugate acid R^1H^+ which proceeds to react slowly with R^2 to the reaction product; for example, in the acid catalysed aldol reaction.

General Catalysis

In general acid catalysis all species capable of donating protons contribute to reaction rate acceleration. The strongest acids are most effective. Reactions in which proton transfer is rate-determining exhibit general acid catalysis, for example diazonium coupling reactions.

$$\text{rate} = -\frac{d[R^1]}{dt} = k_1[SH+][R^1][R^2] + k_2[A^1H][R^1][R^2] + k_3[A^2H][R^1][R^2] + \ldots$$

When keeping the pH at a constant level but changing the buffer concentration a change in rate signals a general acid catalysis. A constant rate is evidence for a specific acid catalyst. When reactions are conducted in nonpolar media, this kind of catalysis is important because the acid is often not ionized. Enzymes catalyze reactions using general-acid and general-base catalysis.

Lewis Acid Catalysis

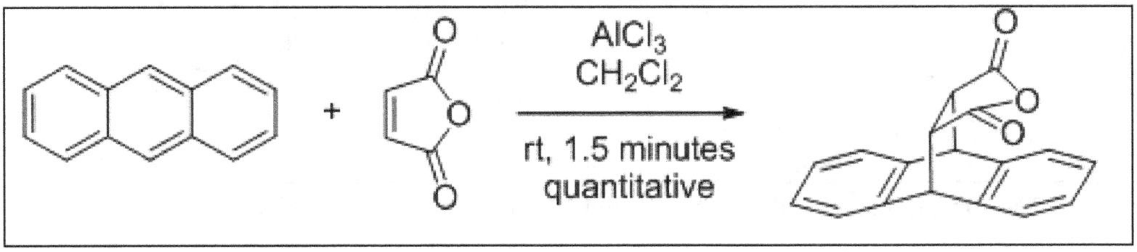

The first Lewis acid-catalyzed Diels–Alder reaction.

In Lewis acid catalysis of organic reactions, a metal-based Lewis acid acts as an electron pair acceptor to increase the reactivity of a substrate. Common Lewis acid catalysts are based on main group metals such as aluminum, boron, silicon, and tin, as well as many early (titanium, zirconium) and late (iron, copper, zinc) d-block metals. The metal atom forms an adduct with a lone-pair bearing electronegative atom in the substrate, such as oxygen (both sp^2 or sp^3), nitrogen, sulfur, and halogens. The complexation has partial charge-transfer character and makes the lone-pair donor effectively more electronegative, activating the substrate toward nucleophilic attack, heterolytic bond cleavage, or cycloaddition with 1,3-dienes and 1,3-dipoles.

Many classical reactions involving carbon–carbon or carbon–heteroatom bond formation can be catalyzed by Lewis acids. Examples include the Friedel-Crafts reaction, the aldol reaction, and various pericyclic processes that proceed slowly at room temperature, such as the Diels-Alder reaction and the ene reaction. In addition to accelerating the reactions, Lewis acid catalysts are able to impose regioselectivity and stereoselectivity in many cases.

Early developments in Lewis acid reagents focused on easily available compounds such as $TiCl_4$, BF_3, $SnCl_4$, and $AlCl_3$. Over the years, versatile catalysts bearing ligands designed for specific applications have facilitated improvement in both reactivity and selectivity of Lewis acid-catalyzed reactions. More recently, Lewis acid catalysts with chiral ligands have become an important class of tools for asymmetric catalysis.

Challenges in the development of Lewis acid catalysis include inefficient catalyst turnover (caused by catalyst affinity for the product) and the frequent requirement of two-point binding for stereoselectivity, which often necessitates the use of auxiliary groups.

Mechanism

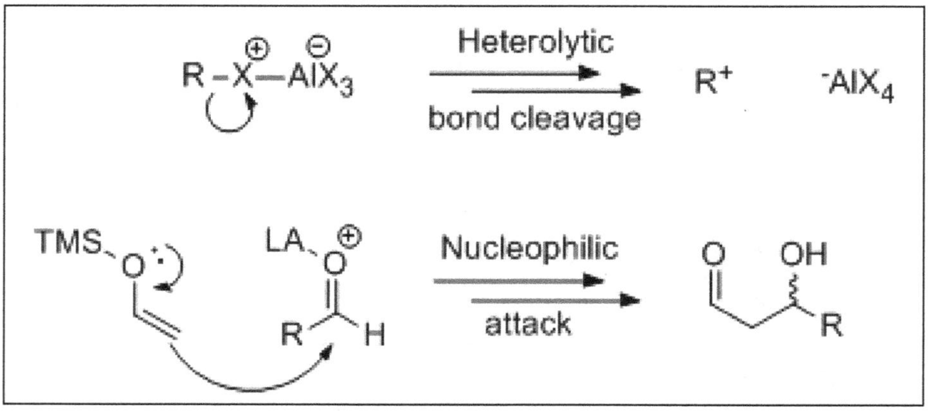

Two common modes of Lewis acid catalysis in reactions with polar mechanisms.

In reactions with polar mechanisms, Lewis acid catalysis often involves binding of the catalyst to Lewis basic heteroatoms and withdrawing electron density, which in turn facilitates heterolytic bond cleavage (in the case of Friedel-Crafts reaction) or directly activates the substrate toward nucleophilic attack (in the case of carbonyl addition reactions). The dichotomy can have important consequences in some reactions, as in the case of Lewis acid-promoted acetal substitution reactions, where the S_N1 and S_N2 mechanisms shown below may give different stereochemical outcomes. Studying the product ratio in a bicyclic system, Denmark and colleagues showed that both mechanisms could be operative depending on the denticity of the Lewis acid and the identity of the R' group.

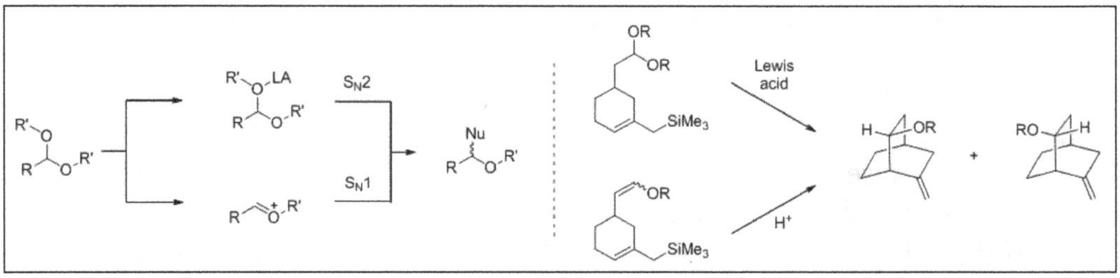

(Left) Lewis acid-promoted acetal substitution may proceed via either the S_N1 or the S_N2 mechanism. (Right) Denmark's model system for acetal substitution mechanism. If the S_N1 mechanism is operative for the acetal substitution, the two reactions shown here should proceed via the same oxocarbenium ion and give similar stereochemical outcomes. Results indicate that the mechanism varies depending on the Lewis acid and the R group.

In Diels-Alder and 1,3-dipolar cycloaddition reactions, Lewis acids lower the LUMO energy of the dienophile or dipolarphile, respectively, making it more reactive toward the diene or the dipole.

Lewis Acid Catalysis with Carbonyl-containing Substrates

Among the types of reactions that can be catalyzed by Lewis acids, those with carbonyl-containing substrates have received the greatest amount of attention. The first major discovery in this area was in 1960, when Yates and Eaton reported the significant acceleration of the Diels-Alder reaction by $AlCl_3$ when maleic anhydride is the dienophile. Early theoretical studies that depended on frontier orbital analysis established that Lewis acid catalysis operates via lowering of the dienophile's LUMO energy, which is still the accepted rationalization. The concept of lowered LUMO energy is also used to explain the dramatically enhanced electrophilic reactivity of carbonyl compounds (whose LUMO is the C-O π^* orbital) towards mild nucleophilic reagents, as in the cases of the Mukaiyama aldol reaction and Sakurai reaction.

In addition to rate acceleration, Lewis acid-catalyzed reactions sometimes exhibit enhanced stereoselectivity, which stimulated the development of stereoinduction models. The models have their roots in knowledge of the structures of Lewis acid-carbonyl complexes which, through decades of research in theoretical calculations, NMR spectroscopy, and X-ray crystallography, were fairly firmly established in the early 1990s:

- σ-Complexation: The complex in which the Lewis acid interacts with the carbonyl compound through a σ-bond with the oxygen lone pair is both thermodynamically favored and catalytically relevant.
- Bent geometry: The metal-oxygen-carbon bond angle is less than 180°, and the metal is syn to the smaller substituent, unless influenced by a chelating group on the larger substituent.
- An s-trans preference for α,β-unsaturated compounds.

Addition and Conjugate Addition to Carbonyl Compounds

The Mukaiyama aldol reaction and the Sakurai reaction refer to the addition of silyl enol ethers and allylsilanes to carbonyl compounds, respectively. Only under Lewis acid catalysis do these reactions occur under synthetically useful conditions. Acyclic transition states are believed to be operating in both reactions for either 1,2- or 1,4-addition, and steric factors control stereoselectivity. This is in contrast with the rigid Zimmerman-Traxler cyclic transition state that has been widely accepted for the aldol reaction with lithium, boron, and titanium enolates. As a consequence, the double bond geometry in the silyl enol ether or allylsilane does not translate well into product stereochemistry. A model for the Sakurai 1,2-addition, proposed by Kumada, is presented in the scheme below; the syn diastereomer is predominant when the (E) silane is used, and also slightly favored when the (Z) silane is used. A similar analysis by Heathcock explains the fact that, with simple substrates, there is essentially no diastereoselectivity for the intermolecular Mukaiyama aldol reaction.

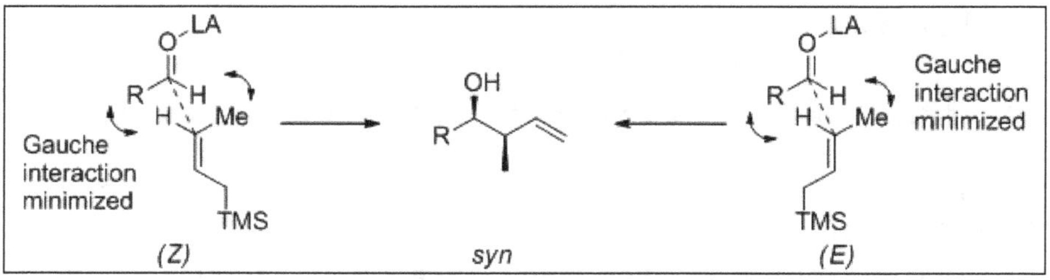

Open transition state model for Sakurai reaction.

The Lewis acid catalyst plays a role in stereoselectivity when the aldehyde can chelate onto the metal center and form a rigid cyclic intermediate. The stereochemical outcome is then consistent with approach of the nucleophile anti to the more bulky substituent on the ring.

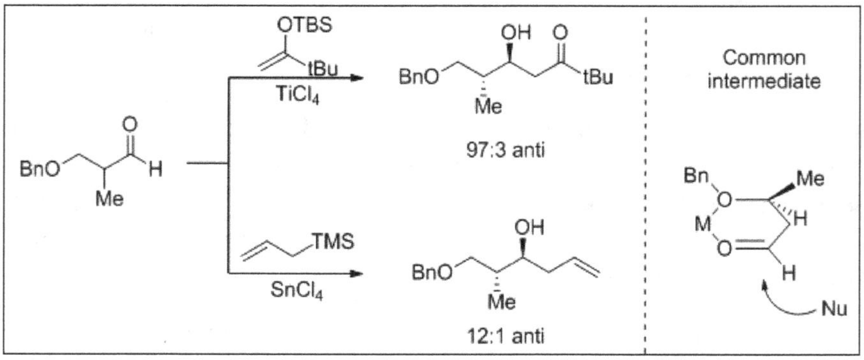

Diels-Alder Reaction

Lewis acids such as $ZnCl_2$, BF_3, $SnCl_4$, $AlCl_3$, and $MeAlCl_2$ can catalyze both normal and inverse electron demand Diels-Alder reactions. The enhancement in rate is often dramatic, and regioselectivity towards ortho- or para-like products is often improved, as shown in the reaction between isoprene and methyl acrylate.

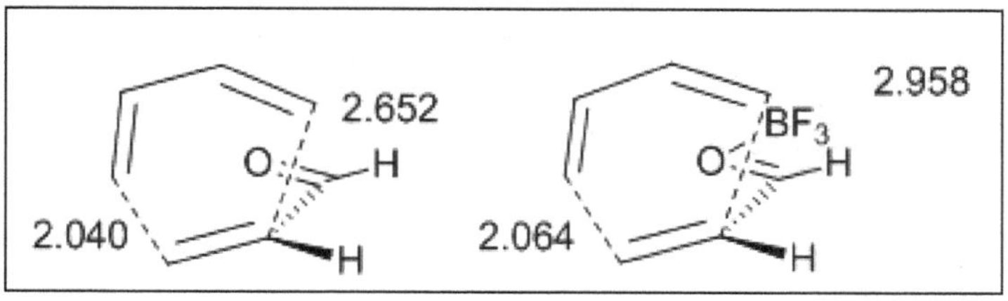

The transition state geometries of Diels-Alder reaction under thermal (left) and BF_3-catalyzed conditions (right). The lengths of the forming bonds (in angstroms) are shown, indicating a more asynchronous transition state for the catalyzed reaction.

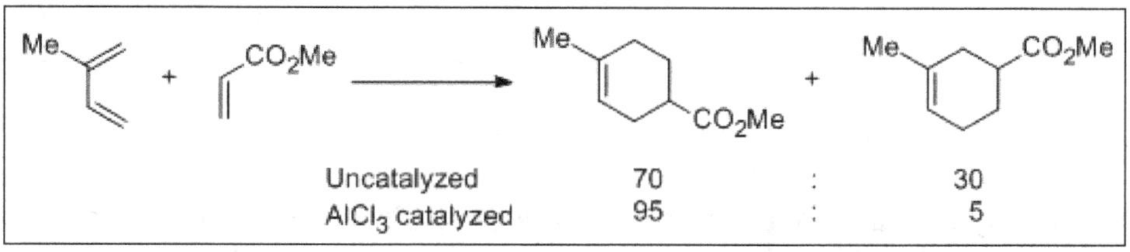

The catalyzed Diels-Alder reaction is believed to be concerted. A computational study at the B3LYP/6-31G(d) level has shown, however, that the transition state of the BF_3-catalyzed Diels-Alder reaction between propenal and 1,3-butadiene is more asynchronous than that of the thermal reaction – the bond further from the carbonyl group is formed ahead of the other bond.

Ene Reaction

The carbonyl-ene reaction is almost always catalyzed by Lewis acids in synthetic applications. A stepwise or a largely asynchronous mechanism has been proposed for the catalyzed reaction based on kinetic isotope effect studies. Nonetheless, cyclic transition states are frequently invoked to interpret diastereoselectivity.

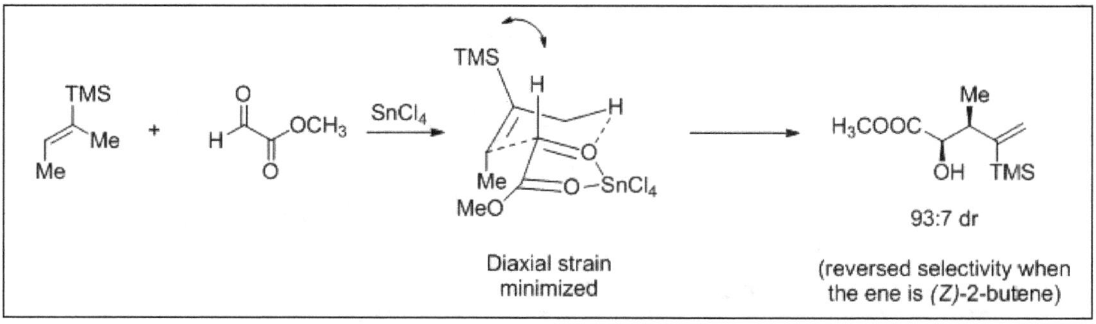

However, the same group carried out HF/6-31G* calculations on tin or aluminum Lewis acid-catalyzed ene reactions. Citing that methyl gloxylate chelates tin Lewis acids but not aluminum ones, they invoked an early, envelope-like transition state and rationalized the divergent stereochemical outcome of the ene reaction between (E)-2-butene and methyl glyoxylate.

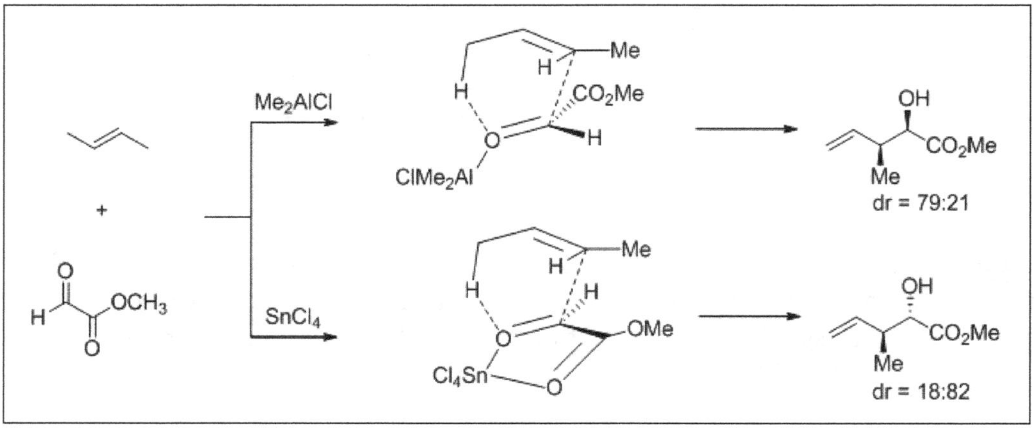

Application in Synthesis

Lewis-acid catalyzed carbonyl addition reactions are routinely used to form carbon–carbon bonds in natural product synthesis. The first two reactions shown below are from the syntheses of (+)-lycoflexine and zaragozic acid C, respectively, which are direct applications of Sakurai and Mukaiyama reactions. The third reaction, en route to (+)-fawcettimine, is a Lewis-acid catalyzed cyclopropane opening that is analogous to a Mukaiyama-Michael reaction.

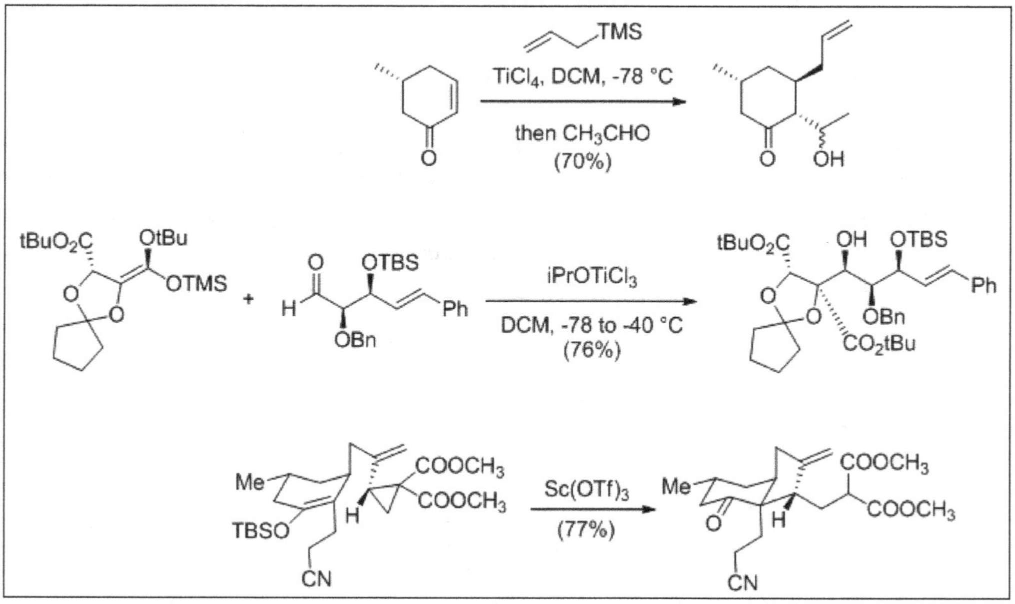

The Diels-Alder reaction catalyzed or promoted by Lewis acids is a powerful and widely used method in natural product synthesis to attain scaffold complexity in a single step with stereochemical control. The two reaction shown below are an intramolecular Diels-Alder reaction towards (–)-fusarisetin A and an intermolecular hetero-Diels-Alder reaction towards (–)-epibatidine, respectively.

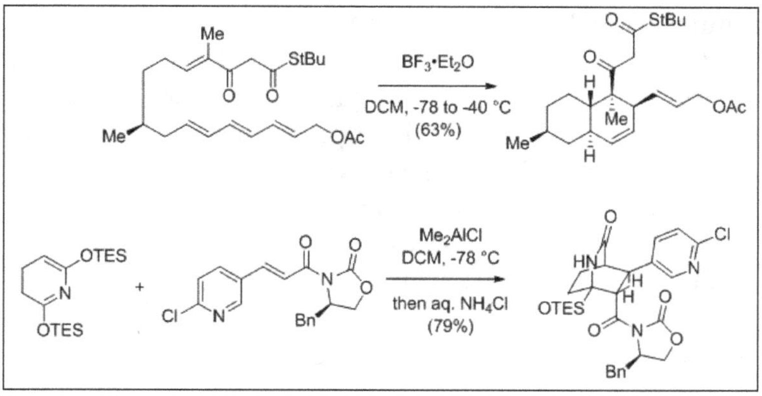

Friedel–Crafts and Related Reactions

In Friedel–Crafts alkylation, a Lewis acid – usually a simple metal halide salt – promotes heterolytic cleavage of a carbon–halogen bond in an alkyl halide and generates a carbocation, which undergoes electrophilic aromatic substitution. Although vastly useful in synthesis, the reaction often suffers from side reactions that arise from carbocation rearrangement, alkyl migration, and over-alkylation. Similarly, in Friedel-Crafts acylation, a Lewis acid assists in the generation of an acylium ion from an acid chloride (or occasionally acid anhydride). Although the acylium ion is often assumed to be the active intermediate, there is evidence that the protonated acylium dication is the active electrophile that undergoes subsequent electrophilic aromatic substitution.

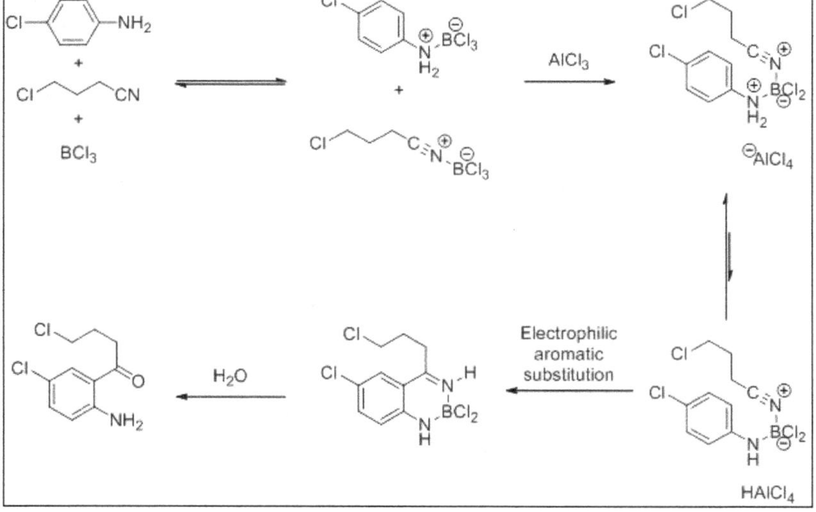

Important variants of the Friedel–Crafts reaction include chloromethylation (with formaldehyde and HCl), formylation (with HCl and CO or CN^-), and acylation with a nitrile as the acyl source. The nitrile-based acylation is particularly useful because it allows direct ortho-acylation of aniline without protecting the amine group. A combination of a weak and a strong Lewis acid is necessary for the reaction to proceed. Guided by this mechanism, and equipped with knowledge that gallium trihalides are among the strongest Lewis acids, process chemists at Merck were able to develop highly efficient conditions for this condition towards a drug candidate.

Asymmetric Lewis Acid Catalysis

Common Chiral Ligands

Asymmetric catalysis by Lewis acids rely on catalysts with chiral ligands coordinated to the metal center. Over the years, a small number of chiral ligand scaffolds have stood out as having "privileged" catalytic properties suitable for a wide range of applications, often of unrelated mechanisms. Current research efforts in asymmetric Lewis acid catalysis mostly utilize or modify those ligands rather than create new scaffolds de novo. The "privileged" scaffolds share a few common features, including chemical stability and relative ease of elaboration. The majority of the scaffolds are multidentate. Most of them also have high scaffold rigidity within the ligand.

Bisoxazolines (BOX)

Most common chiral bisoxazoline (BOX) ligands consist of two identical chiral oxazoline moieties, substituted by a bulky group at the 4-positions, joined by a linker. The ligand is bidentate when the linker is a single carbon unit, but is tridentate (usually meridial) when the linker bears an additional coordinating atom, such as a pyridine nitrogen in the case of PyBOX ligands. The impact of ligand denticity and active intermediate geometry on the stereochemical outcome has been thoroughly reviewed.

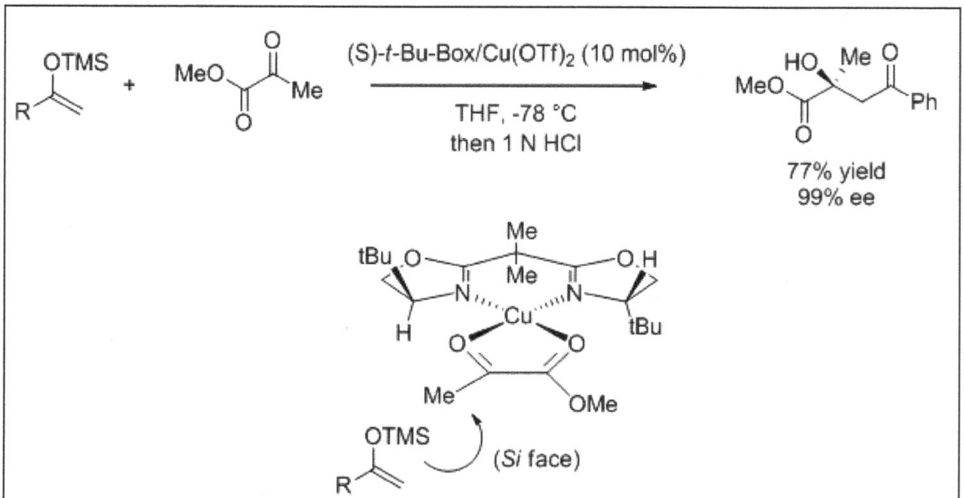

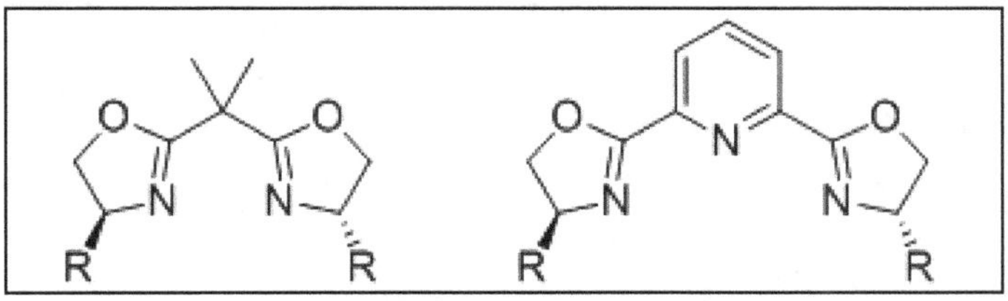

Generic structure of BOX (left) and PyBOX (right) ligands.

Many bidentate BOX-based Lewis acid-catalyzed reactions are based on copper(II) catalysts with substrates that are suitable for two-point binding. The stereochemical outcome is consistent with a twisted square planar intermediate that was proposed based on related crystal structures. The substituent at the oxazoline's 4-position blocks one enantiotopic face of the substrate, leading to enantioselectivity. This is demonstrated in the following aldol-type reaction, but is applicable to a wide variety of reactions such as Mannich-type reactions, ene reaction, Michael addition, Nazarov cyclization, and hetero-Diels-Alder reaction.

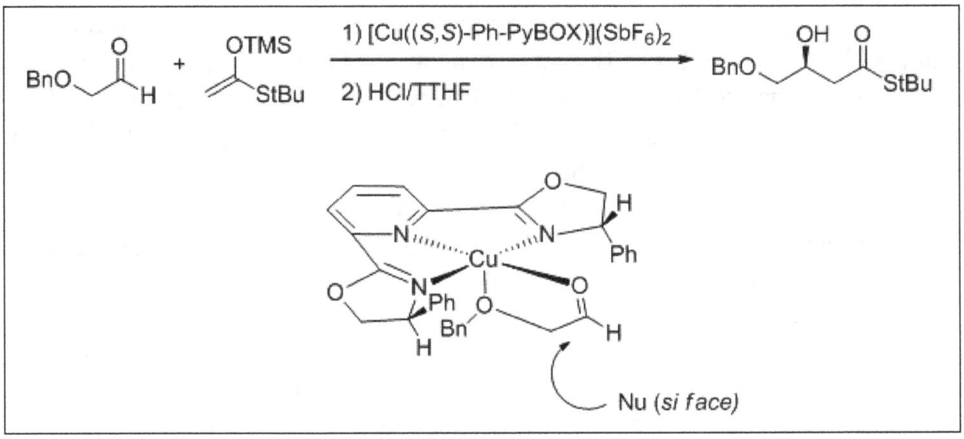

On the other hand, two-point binding on a Lewis acid bearing the meridially tridentate PyBOX ligand would result in a square pyramidal complex. The stereochemical outcome is consistent with the carbonyl oxygen binding equatorially and the ether oxygen binding axially.

Binap

Developed by Noyori, BINAP (2,2'-diphenylphosphino-1,1'-binaphthyl) is a family of chiral diphosphine ligands featuring two triarylphosphine moieties installed on a binaphthalene backbone. BINAP chelates onto a metal (usually a late transition metal) to form a C_2-symmetric complex. In the structure of an (R)-BINAP ruthenium complex, among the four remaining coordination sites on an octahedral metal center, the two equatorial sites (purple) are strongly influenced by the equatorial phenyl groups, while the two axial sites (green) are influenced by the axial phenyl groups.

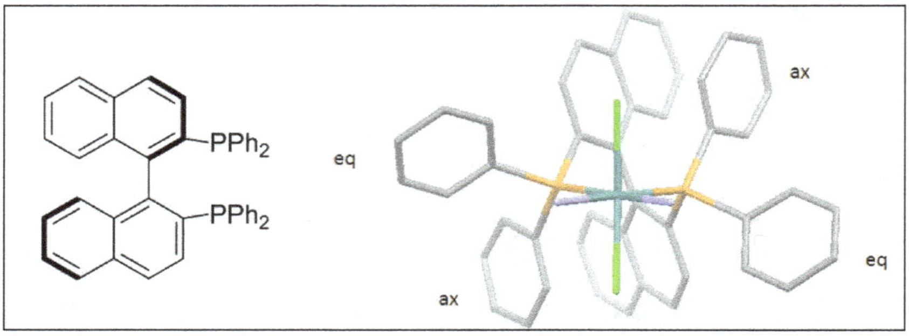

Left: Structure of (R)-BINAP. Right: Structure of an (R)-BINAP ruthenium complex, highlighting the equatorial (purple) and axial (green) coordination sites, and the equatorial and axial phenyl groups that enforce the asymmetric environment for incoming ligands.

Based on the structure, models for the observed enantioselectivity in many BINAP-based Lewis acid-catalyzed reactions have been proposed. For example, in the palladium-catalyzed enantioselective Diels-Alder reaction shown below, the dienophile is thought to coordinate the metal center at the equatorial sites. Thus the equatorial phenyl group on phosphorus obstructs the Si-face, resulting in excellent enantioselectivity. A very similar model was used to rationalize the outcome of a nickel-catalyzed asymmetric enolate alkylation reaction, where the substrate also bears an auxiliary that allows it to chelate onto the metal. On the other hand, a copper(I)-catalyzed hetero-ene reaction is thought to proceed through a tetrahedral intermediate, offering an alternative mode of stereoinduction by changing the metal center.

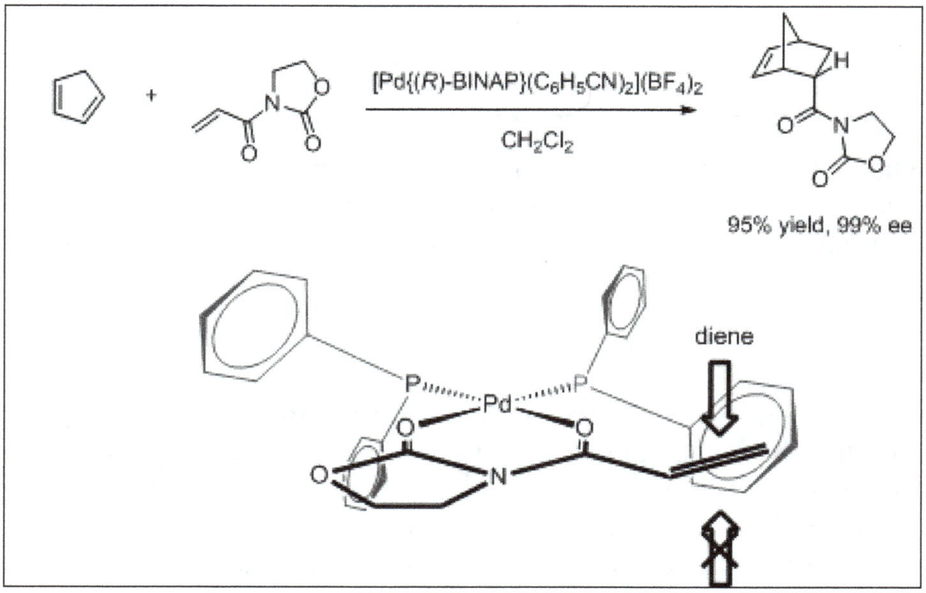

A BINAP-palladium-catalyzed asymmetric Diels-Alder reaction. In the model for the reaction intermediate, the binaphthyl rings are omitted for clarity.

Binol

BINOL (1,1'-binaphthyl-2,2'-diol) is usually used in conjunction with oxophilic Lewis acidic metals such as aluminum, titanium, zirconium, and various rare earth metals. In cases where BINOL itself does not provide ideal enantioselective control, it can be readily elaborated by substitution at the 3,3' positions (via lithiation) and 6,6' positions (via the 6,6'-dibromide compound prepared by electrophilic aromatic substitution) to modulate steric bulk and electronic properties. For instance, aluminum catalysts based on bulky 3,3'-disilyl substituted BINOL have been developed as early examples of catalytic asymmetric hetero-Diels-Alder reaction and Claisen rearrangement, while introduction of electron-withdrawing groups at the 6,6'-positions was crucial for increasing the Lewis acidity, and hence catalytic activity, of zirconium(IV) catalysts toward a Mannich-type reaction. To date, however, no model for the crucial factors governing BINOL-directed stereoinduction has been generally accepted.

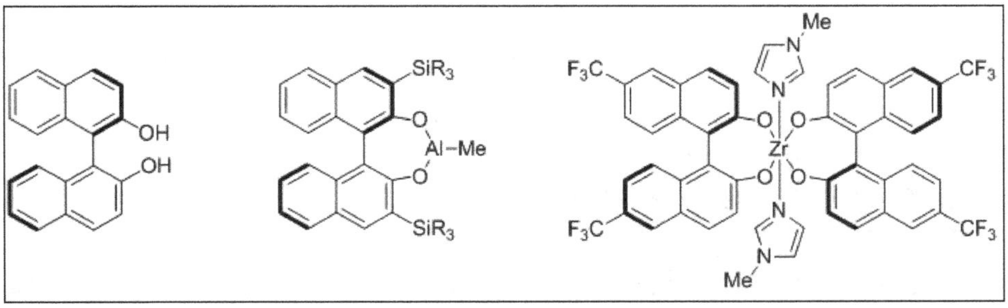

Left: (R)-BINOL. Center: Aluminum catalyst based on bulky 3,3'-disilyl substituted BINOL. Right: Zirconium catalyst based on BINOL substituted at 6 and 6' positions by electron-withdrawing trifluoromethyl group.

Taddol

TADDOL stands for tetraaryl-1,3-dioxolane-4,5-dimethanol.

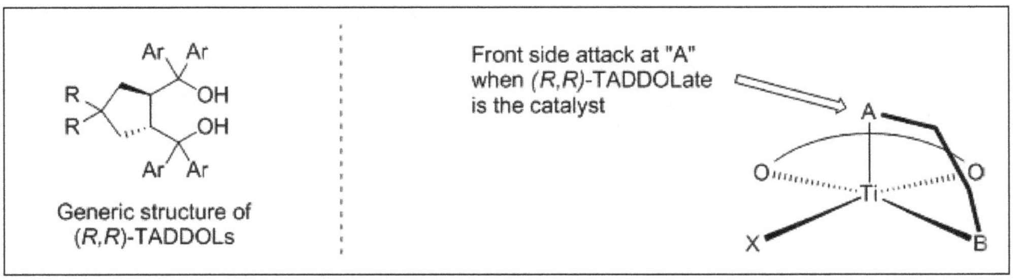

Applications

Lewis acid catalysis has been used in the asymmetry-setting step for the syntheses of many natural products. The first reaction shown below, from the synthesis of taxane skeleton, uses a copper-based catalyst supported by a chiral phosphoramidite ligand for a conjugate carbonyl addition reaction. The second reaction, from the synthesis of ent-hyperforin, uses an iron-PyBOX catalyst for an asymmetric Diels-Alder reaction.

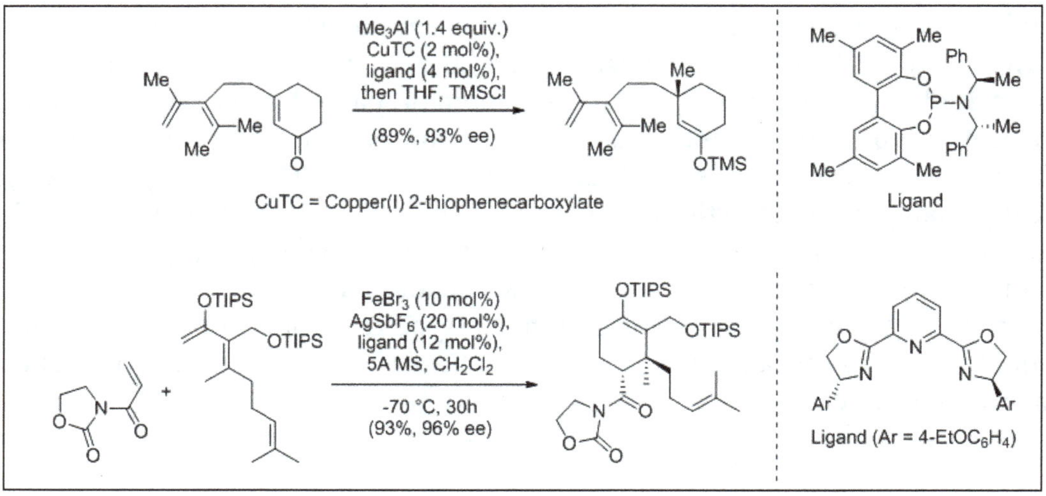

Heterogenous Catalysis

Heterogeneous catalysis is the type of catalysis where the phase of the catalyst differs from the phase of the reactants or products. Contrasts with homogeneous catalysis where the reactants, products and catalyst exist in the same phase. Phase distinguishes between not only solid, liquid, and gas components, but also immiscible mixtures (e.g. oil and water), or anywhere an interface is present. Catalysts are useful because they increase the rate of a reaction without themselves being consumed and are therefore reusable.

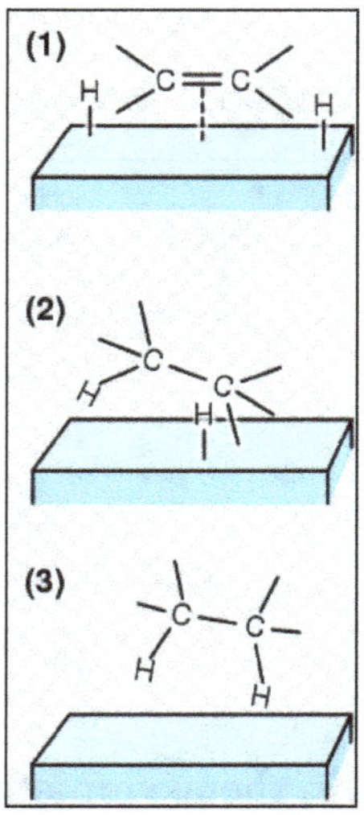

Hydrogenation of ethene on a catalytic solid surface: (1) Adsorption (2) Reaction (3) Desorption.

Heterogeneous catalysis typically involves solid phase catalysts and gas phase reactants. In this case, there is a cycle of molecular adsorption, reaction, and desorption occurring at the catalyst surface. Thermodynamics, mass transfer, and heat transfer influence the rate (kinetics) of reaction.

Heterogeneous catalysis is very important because it enables faster, large-scale production and the selective product formation. Approximately 35% of the world's GDP is influenced by catalysis. The production of 90% of chemicals (by volume) is assisted by solid catalysts. The chemical and energy industries rely heavily on heterogeneous catalysis. For example, the Haber-Bosch process uses metal-based catalysts in the synthesis of ammonia, an important component in fertilizer.

Adsorption

Adsorption is an essential step in heterogeneous catalysis. Adsorption is the process by which a gas (or solution) phase molecule (the adsorbate) binds to solid (or liquid) surface atoms (the adsorbent). The reverse of adsorption is desorption, the adsorbate splitting from adsorbent. In a reaction facilitated by heterogeneous catalysis, the catalyst is the adsorbent and the reactants are the adsorbate.

Types of Adsorption

Two types of adsorption are recognized: physisorption, weakly bound adsorption, and chemisorption, strongly bound adsorption. Many processes in heterogeneous catalysis lie between the two extremes. The Lennard-Jones model provides a basic framework for predicting molecular interactions as a function of atomic separation.

Physisorption

In physisorption, a molecule becomes attracted to the surface atoms via van der Waals forces. These include dipole-dipole interactions, induced dipole interactions, and London dispersion forces. No chemical bonds are formed between adsorbate and adsorbent, and their electronic states remain relatively unperturbed. Typical energies for physisorption are from 3 to 10 kcal/mol. In heterogeneous catalysis, when a reactant molecule physisorbs to a catalyst, it is commonly said to be in a precursor state, an intermediate energy state before chemisorption, a more strongly bound adsorption. From the precursor state, a molecule can either undergo chemisorption, desorption, or migration across the surface. The nature of the precursor state can influence the reaction kinetics.

Chemisorption

When a molecule approaches close enough to surface atoms such that their electron clouds overlap, chemisorption can occur. In chemisorption, the adsorbate and adsorbent share electrons signifying the formation of chemical bonds. Typical energies for chemisorption range from 20 to 100 kcal/mol. Two cases of chemisorption are:

- Molecular adsorption: The adsorbate remains intact. An example is alkene binding by platinum.
- Dissociation adsorption: One or more bonds break concomitantly with adsorption. In this case, the barrier to dissociation affects the rate of adsorption. An example of this the binding of H_2 to a metal catalyst, where the H-H bond is broken upon adsorption.

Surface Reactions

Most metal surface reactions occur by chain propagation in which catalytic intermediates are cyclically produced and consumed. Two main mechanisms for surface reactions can be described for A + B → C.

- Langmuir-Hinshelwood mechanism: The reactant molecules, A and B, both adsorb to the catalytic surface. While adsorbed to the surface, they combine to form product C, which then desorbs.
- Eley-Rideal mechanism: One reactant molecule, A, adsorbs to the catalytic surface. Without adsorbing, B reacts with absorbed A to form C, that then desorbs from the surface.

Most heterogeneously catalyzed reactions are described by the Langmuir-Hinshelwood model.

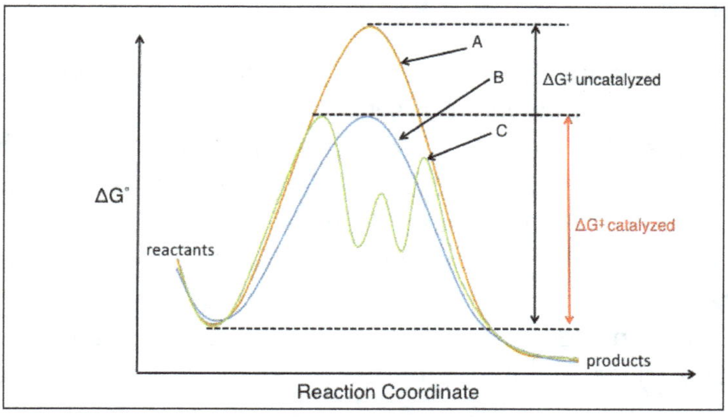

Reaction Coordinate: (A) Uncatalyzed (B) Catalyzed (C) Catalyzed with discreet intermediates (transition states).

In heterogeneous catalysis, reactants diffuse from the bulk fluid phase to adsorb to the catalyst surface. The adsorption site is not always an active catalyst site, so reactant molecules must migrate across the surface to an active site. At the active site, reactant molecules will react to form product molecule(s) by following a more energetically facile path through catalytic intermediates. The product molecules then desorb from the surface and diffuse away. The catalyst itself remains intact and free to mediate further reactions. Transport phenomena such as heat and mass transfer, also play a role in the observed reaction rate.

Catalyst Design

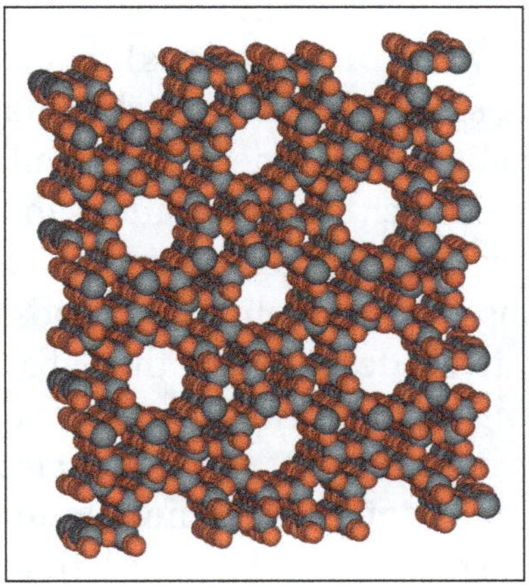

Zeolite structure: A common catalyst support material in hydrocracking. Also acts as a catalyst in hydrocarbon alkylation and isomerization.

Catalysts are not active towards reactants across their entire surface; only specific locations possess catalytic activity, called active sites. The surface area of a solid catalyst has a strong influence on the number of available active sites. In industrial practice, solid catalysts are often porous to maximize surface area, commonly achieving 50–400 m^2/g. Some mesoporous silicates, such as the MCM-41, have surface areas greater than 1000 m^2/g. Porous materials are cost effective due to their high surface area-to-mass ratio and enhanced catalytic activity.

In many cases, a solid catalyst is dispersed on a supporting material to increase surface area (spread the number of active sites) and provide stability. Usually catalyst supports are inert, high melting point materials, but they can also be catalytic themselves. Most catalyst supports are porous (frequently carbon, silica, zeolite, or alumina-based) and chosen for their high surface area-to-mass ratio. For a given reaction, porous supports must be selected such that reactants and products can enter and exit the material.

Often, substances are intentionally added to the reaction feed or on the catalyst to influence catalytic activity, selectivity, and/or stability. These compounds are called promoters. For example, alumina (Al_2O_3) is added during ammonia synthesis to providing greater stability by slowing sintering processes on the Fe-catalyst.

Sabatier principle can be considered as one of the cornerstones of modern theory of catalysis. Sabatier principle states that the surface-adsorbates interaction has to be an optimal amount and not too weak to be inert toward the reactants and not too strong to poison the surface and avoid desorption of the products. The statement that the surface-adsorbate interaction has to be an optimum, is a qualitative one. Usually the number of adsorbates and transition states associated with a chemical reaction is a large number, thus the optimum has to be found in a many-dimensional space. Catalyst design in such a many-dimensional space is not a computationally viable task. Additionally, such optimization process would be far from intuitive. Scaling relations are used to decrease the dimensionality of the space of catalyst design. Such relations are correlations among adsorbates binding energies (or among adsorbate binding energies and transition states also known as BEP relations) that are "similar enough" e.g., OH versus OOH scaling. Applying scaling relations to the catalyst design problems greatly reduces the space dimensionality (sometimes to as small as 1 or 2). One can also use micro-kinetic modeling based on such scaling relations to take into account the kinetics associated with adsorption, reaction and desorption of molecules under specific pressure or temperature conditions. Such modeling then leads to well-known volcano-plots at which the optimum qualitatively described by the Sabatier principle is referred to as the "top of the volcano". Scaling relations can be used not only to connect the energetics of radical surface-adsorbed groups (e.g., O*,OH*), but also to connect the energetics of closed-shell molecules among each other or to the counterpart radical adsorbates. The correlations which are manifested in the scaling relations confine the catalyst design space, preventing one from reaching the "top of the volcano".

Breaking scaling relations can refer to either designing surfaces or motifs that do not follow a scaling relation, or ones that follow a different scaling relation (than the usual relation for the associated adsorbates) in the right direction: one that can get us closer to the top of the reactivity volcano. In addition to studying catalytic reactivity, scaling relations can be used to study and screen materials for selectivity toward a special product. There are special combination of binding energies that favor specific products over the others. Sometimes a set of binding energies that can change the selectivity toward a specific product "scale" with each other, thus to improve the selectivity one has to break some scaling relations. An example of this is the scaling between methane and methanol oxidative activation energies that leads to the lack of selectivity in direct conversion of methane to methanol.

Catalyst Deactivation

Catalyst deactivation is defined as a loss in catalytic activity and/or selectivity over time. Substances that decrease reaction rate are called poisons. Poisons chemisorb to catalyst surface and reduce the number of available active sites for reactant molecules to bind to. Common poisons include Group V, VI, and VII elements (e.g. S, O, P, Cl), some toxic metals (e.g. As, Pb), and adsorbing species with multiple bonds (e.g. CO, unsaturated hydrocarbons). For example, sulfur disrupts the production of methanol by poisoning the Cu/ZnO catalyst. Substances that increase reaction rate are called promoters. For example, the presence of alkali metals in ammonia synthesis increases the rate of N_2 dissociation.

The presence of poisons and promoters can alter the activation energy of the rate-limiting step and affect a catalyst's selectivity for the formation of certain products. Depending on the amount, a substance can be favorable or unfavorable for a chemical process. For example, in the production of ethylene, a small amount of chemisorbed chlorine will act as a promoter by improving Ag-catalyst selectivity towards ethylene over CO_2, while too much chlorine will act as a poison.

Other mechanisms for catalyst deactivation include:

- Sintering: When heated, dispersed catalytic metal particles can migrate across the support surface and form crystals. This results in a reduction of catalyst surface area.
- Fouling: The deposition of materials from the fluid phase onto the solid phase catalyst and/or support surfaces. This results in active site and/or pore blockage.
- Coking: The deposition of heavy, carbon-rich solids onto surfaces due to the decomposition of hydrocarbons.
- Vapor-solid reactions: Formation of an inactive surface layer and/or formation of a volatile compound that exits the reactor. This results in a loss of surface area and/or catalyst material.
- Solid-state transformation: Solid-state diffusion of catalyst support atoms to the surface followed by a reaction that forms an inactive phase. This results in a loss of catalyst surface area.
- Erosion: Continual attrition of catalyst material common in fluidized-bed reactors. This results in a loss of catalyst material.

In industry, catalyst deactivation costs billions every year due to process shutdown and catalyst replacement.

Industrial Examples

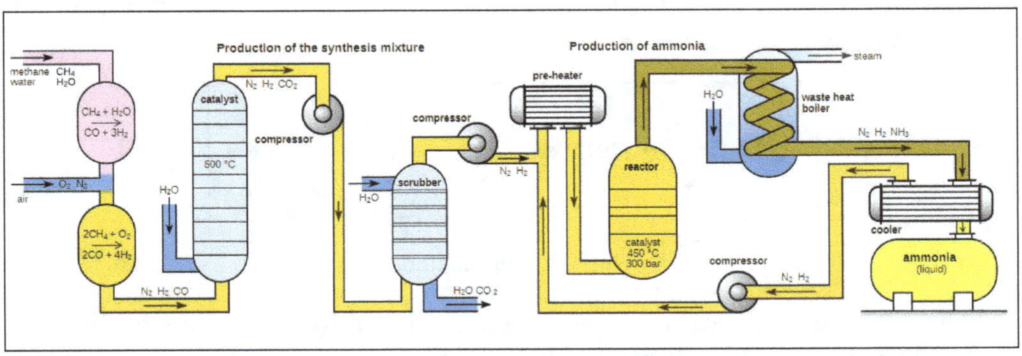

Process flow diagram illustrating the use of catalysis in the synthesis of ammonia (NH_3).

In industry, many design variables must be considered including reactor design. The conventional heterogeneous catalysis reactors include batch, continuous, and fluidized-bed reactors, while more recent setups include fixed-bed, microchannel, and multi-functional reactors. Other variables to consider are reactor dimensions, surface area, catalyst type, catalyst support, as well as reactor operating conditions such as temperature, pressure, and reactant concentrations.

Table: Some large-scale industrial processes incorporating heterogeneous catalysts.

Process	Reactants, Product(s)	Catalyst	Comment
Sulfuric acid synthesis (Contact process)	$SO_2 + O_2$, SO_3	Vanadium oxides	Hydration of SO_3 gives H_2SO_4
Ammonia synthesis (Haber–Bosch process)	$N_2 + H_2$, NH_3	Iron oxides on alumina(Al_2O_3)	Consumes 1% of world's industrial energy budget
Nitric acid synthesis (Ostwald process)	$NH_3 + O_2$, HNO_3	Unsupported Pt-Rh gauze	Direct routes from N_2 are uneconomical
Hydrogen production by Steam reforming	$CH_4 + H_2O$, $H_2 + CO_2$	Nickel or K_2O	Greener routes to H2 by water splitting actively sought
Ethylene oxide synthesis	$C_2H_4 + O_2$, C_2H_4O	Silver on alumina, with many promoters	Poorly applicable to other alkenes
Hydrogen cyanide synthesis (Andrussov oxidation)	$NH_3 + O_2 + CH_4$, HCN	Pt-Rh	Related ammoxidation process converts hydrocarbons to nitriles
Olefin polymerization Ziegler–Natta polymerization	Propylene, polypropylene	TiCl3 on MgCl2	Many variations exist, including some homogeneous examples
Desulfurization of petroleum (hydrodesulfurization)	$H_2 + R_2S$ (idealized organosulfur impurity), $RH + H_2S$	Mo-Co on alumina	Produces low-sulfur hydrocarbons, sulfur recovered via the Claus process

Other Examples:

- Reduction of nitriles in the synthesis of phenethylamine with Raney nickel catalyst and hydrogen in ammonia:

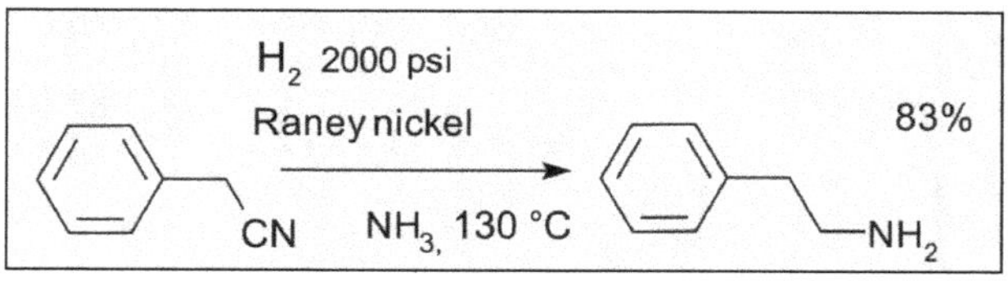

Nitrile hydrogenation.

- The cracking, isomerisation, and reformation of hydrocarbons to form appropriate and useful blends of petrol.

- In automobiles, catalytic converters are used to catalyze three main reactions:
  - The oxidation of carbon monoxide to carbon dioxide:

    $2CO(g) + O_2(g) \rightarrow 2CO_2(g)$
  - The reduction of nitrogen monoxide back to nitrogen:

    $2NO(g) + 2CO(g) \rightarrow N_2(g) + 2CO_2(g)$
  - The oxidation of hydrocarbons to water and carbon dioxide:

    $2\,C_6H_6 + 15\,O_2 \rightarrow 12\,CO_2 + 6\,H_2O$
- This process can occur with any of hydrocarbon, but most commonly is performed with petrol or diesel.
- Asymmetric heterogeneous catalysis facilitates the production of pure enantiomer compounds using chiral heterogeneous catalysts.
- The vast majority of heterogeneous catalysts are based on metals or metal oxides; however, some chemical reactions can be catalyzed by carbon-based materials, e.g., oxidative dehydrogenations or selective oxidations.
  - Ethylbenzene + $1/2\,O_2 \rightarrow$ Styrene + H_2O
  - Acrolein + $1/2\,O_2 \rightarrow$ Acrylic acid

Solid-liquid and liquid-liquid Catalyzed Reactions

Although the majority of heterogeneous catalysts are solids, there are a few variations which are of practical value. For two immiscible solutions (liquids), one carries the catalyst while the other carries the reactant. This set up is the basis of biphasic catalysis as implemented in the industrial production of butyraldehyde by the hydroformylation of propylene.

Reacting phases	Examples given	Comment
Solid + solution	Hydrogenation of fatty acids with nickel	Used for the production of margarine
Immiscible liquid phases	Hydroformylation of propene	Aqueous phase catalyst; reactants and products mainly in non-aqueous phase

Zeolites

Zeolites are microporous, aluminosilicate minerals commonly used as commercial adsorbents and catalysts. The term zeolite was originally coined in 1756 by Swedish mineralogist Axel Fredrik Cronstedt, who observed that rapidly heating the material, believed to have been stilbite, produced large amounts of steam from water that had been adsorbed by the material.

Natural zeolite with a penny for scale.

Zeolites occur naturally but are also produced industrially on a large scale. 245 unique zeolite frameworks have been identified, and over 40 naturally occurring zeolite frameworks are known. Every new zeolite structure that is obtained is examined by the International Zeolite Association Structure Commission and receives a three letter designation.

Properties and Occurrence

Zeolites have a porous structure that can accommodate a wide variety of cations, such as Na^+, K^+, Ca^{2+}, Mg^{2+} and others. These positive ions are rather loosely held and can readily be exchanged for others in a contact solution. Some of the more common mineral zeolites are analcime, chabazite, clinoptilolite, heulandite, natrolite, phillipsite, and stilbite. An example of the mineral formula of a zeolite is $Na_2Al_2Si_3O_{10}{\cdot}2H_2O$, the formula for natrolite. These cation exchanged zeolites possess different acidity and catalyse several acid catalysis.

A form of thomsonite (one of the rarest zeolites).

Natural zeolites form where volcanic rocks and ash layers react with alkaline groundwater. Zeolites also crystallize in post-depositional environments over periods ranging from thousands to millions of years in shallow marine basins. Naturally occurring zeolites are rarely pure and are contaminated to varying degrees by other minerals, metals, quartz, or other zeolites. For this reason, naturally occurring zeolites are excluded from many important commercial applications where uniformity and purity are essential.

Zeolites are the aluminosilicate(AlO_4^{5-} and SiO_4^{4-}) members of the family of microporous solids known as "molecular sieves", and mainly consist of Si, Al, O, and metals including Ti, Sn, Zn, and so on. The term molecular sieve refers to a particular property of these materials, i.e., the ability to selectively sort molecules based primarily on a size exclusion process. This is due to a very regular pore structure of molecular dimensions. The maximum size of the molecular or ionic species that can enter the pores of a zeolite is controlled by the dimensions of the channels. These are conventionally defined by the ring size of the aperture, where, for example, the term "8-ring" refers to a closed loop that is built from eight tetrahedrally coordinated silicon (or aluminium) atoms and 8 oxygen atoms. These rings are not always perfectly symmetrical due to a variety of causes, including strain induced by the bonding between units that are needed to produce the overall structure, or coordination of some of the oxygen atoms of the rings to cations within the structure. Therefore, the pores in many zeolites are not cylindrical.

Zeolites transform to other minerals under weathering, hydrothermal alteration or metamorphic conditions. Some examples:

- The sequence of silica-rich volcanic rocks commonly progresses from:
    - Clay → quartz → mordenite–heulandite → epistilbite → stilbite → thomsonite–mesolite-scolecite → chabazite → calcite.
- The sequence of silica-poor volcanic rocks commonly progresses from:
    - Cowlesite → levyne–offretite → analcime → thomsonite–mesolite-scolecite → chabazite → calcite.

Nanomaterial-based Catalyst

Nanomaterial-based catalysts are usually heterogeneous catalysts broken up into metal nanoparticles in order to enhance the catalytic process. Metal nanoparticles have high surface area, which can increase catalytic activity. Nanoparticle catalysts can be easily separated and recycled. They are typically used under mild conditions to prevent decomposition of the nanoparticles.

Functionalized Nanoparticles

Functionalized metal nanoparticles are more stable toward solvents compared to non-functionalized metal nanoparticles. In liquids, the metal nanoparticles can be affected by van der Waals force. Particle aggregation can sometimes decrease catalytic activity by lowering the surface area. Nanoparticles can also be functionalized with polymers or oligomers to sterically stabilize the nanoparticles by providing a protective layer that prevents the nanoparticles from interacting with each other. Alloys of two metals, called bimetallic nanoparticles, are used to create synergistic effects on catalysis between the two metals.

Potential Applications

Dehalogenation and Hydrogenation

Nanoparticle catalysts are active for the hydrogenolysis of C-Cl bonds such as polychlorinated biphenyls. Another reaction is hydrogenation of halogenated aromatic amines is also important for the synthesis of herbicides and pesticides as well as diesel fuel. In organic chemistry, hydrogenation of a C-Cl bond with deuterium is used to selectively label the aromatic ring for use in experiments dealing with the kinetic isotope effect. Buil et al. created rhodium complexes that generated rhodium nanoparticles. These nanoparticles catalyzed the dehalogenation of aromatic compounds as well as the hydrogenation of benzene to cyclohexane. Polymer-stabilized nanoparticles can also be used for the hydrogenation of cinnamaldehyde and citronellal. Yu et al. found that the ruthenium nanocatalysts are more selective in the hydrogenation of citronellal compared to the traditional catalysts used.

Hydrosilylation Reactions

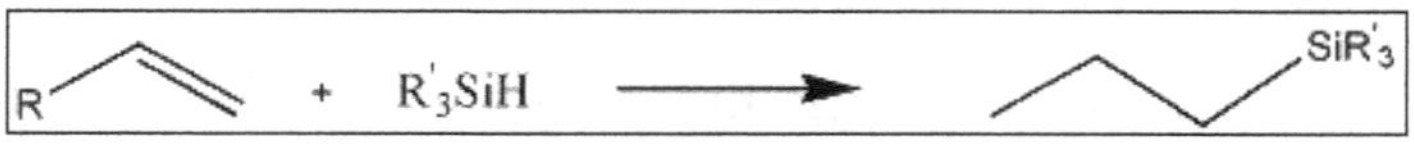

Hydrosilylation reaction.

The Reduction of gold, cobalt, nickel, palladium, or platinum organometallic complexes with silanes produces metal nanoparticle that catalyze the hydrosilylation reaction. BINAP-functionalized palladium nanoparticles and gold nanoparticles have been used for the hydrosilylaytion of styrene under mild conditions; they were found to be more catalytically active and more stable than non-nanoparticle Pd-BINAP complexes. The reaction may also be catalyzed by a nanoparticle that consists of two metals.

Organic Redox Reactions

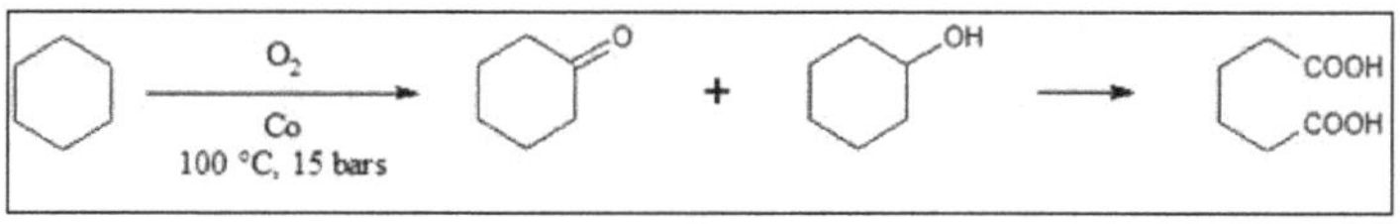

Oxidation reaction of cyclohexane to synthesize adiapic acid.

An oxidation reaction to form adipic acid is shown in figure above and it can be catalyzed by cobalt nanoparticles. This is used in an industrial scale to produce the nylon 6,6 polymer. Other examples of oxidation reactions that are catalyzed by metallic nanoparticles include the oxidation of cyclooctane, the oxidation of ethene, and glucose oxidation.

C-C Coupling Reactions

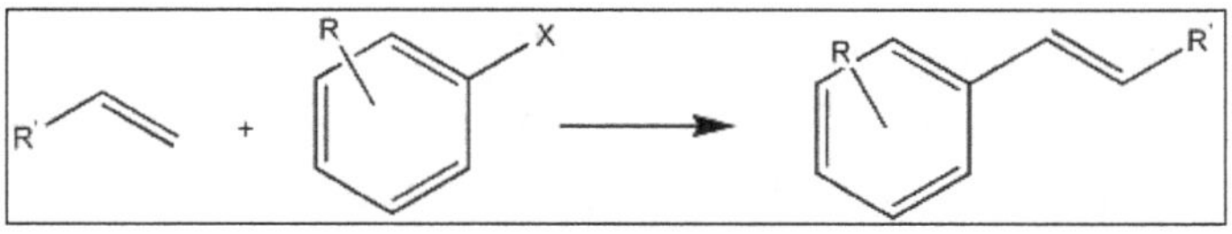

Heck coupling reaction.

Metallic nanoparticles can catalyze C–C coupling reactions such as the hydroformylation of olefins, the synthesis of vitamin E and the Heck coupling and Suzuki coupling reactions.

Palladium nanoparticles were found to efficiently catalyze Heck coupling reactions. It was found that increased electronegativity of the ligands on the palladium nanoparticles increased their catalytic activity.

The compound $Pd_2(dba)_3$ is a source of Pd(0), which is the catalytically active source of palladium used for many reactions, including cross coupling reactions. Pd2(dba)3 was thought to be a homogeneous catalytic precursor.

Alternative Fuels

Iron oxide and cobalt nanoparticles can be loaded onto various surface active materials like alumina to convert gases such as carbon monoxide and hydrogen into liquid hydrocarbon fuels using the Fischer-Tropsch process.

Much research on nanomaterial-based catalysts has to do with maximizing the effectiveness of the catalyst coating in fuel cells. Platinum is currently the most common catalyst for this application, however, it is expensive and rare, so a lot of research has been going into maximizing the catalytic properties of other metals by shrinking them to nanoparticles in the hope that someday they will be an efficient and economic alternative to platinum. Gold nanoparticles also exhibit catalytic properties, despite the fact that bulk gold is unreactive.

Yttrium stabilized zirconium nanoparticles were found to increase the efficiency and reliability of a solid oxide fuel cell. Nanomaterial ruthenium/platinum catalysts could potentially be used to catalyze the purification of hydrogen for hydrogen storage. Palladium nanoparticles can be functionalized with organometallic ligands to catalyze the oxidation of CO and NO to control air pollution in the environment. Carbon nanotube supported catalysts can be used as a cathode catalytic support for fuel cells and metal nanoparticles have been used to catalyze the growth of carbon nanotubes. Platinum-cobalt bimetallic nanoparticles combined with carbon nanotubes are promising candidates for direct methanol fuel cells since they produce a higher stable current electrode.

Medicine

In magnetic chemistry, nanoparticles can be used for catalyst support for medicinal use.

Nanozymes

Besides conventional catalysis, nanomaterials have been explored for mimicking natural enzymes. The nanomaterials with enzyme mimicking activities are termed as nanozymes. Lots of nanomaterials have been used to mimic varieties of natural enzymes, such as oxidase, peroxidase, catalase, SOD, nuclease, etc. The nanozymes have found wide applications in many areas, from biosensing and bioimaging to therapeutics and water treatment.

Nanostructures for Electrocatalysis

Nanocatalysts are of wide interest in fuel cells and electrolyzers, where the catalyst strongly affects efficiency.

- Nanoporous Surfaces: In fuel cells, nanoporous materials are widely used to make cathodes. Porous nanoparticles of platinum have good activity in nanocatalysis but are less stable and their lifetime is short.
- Nanoparticles: One drawback to the use of nanoparticles is their tendency to agglomerate. The problem can be mitigated with the correct catalyst support. Nanoparticles are optimal structures to be used as nanosensors because they can be tuned to detect specific molecules. Examples of Pd nanoparticles electrodeposited on multi-walled carbon nanotubes have shown good activity towards catalysis of cross-coupling reactions.
- Nanowires: Nanowires are very interesting for electrocatalytic purpose because they are easier to produce and the control over their characteristics in the production process is quite precise. Also, nanowires can increase faradaic efficiency due to their spatial extent and thus to greater availability of reactants on the active surface.
- Materials: The nanostructures involved in electrocatalysis processes can be made up of different materials. Through the use of nanostructured materials, electrocatalysts can achieve good physical-chemical stability, high activity, good conductivity and low cost. Metallic nanomaterials are commonly made up of transition metals (mostly iron, cobalt, nickel, palladium, platinum). Multi-metal nanomaterials show new properties due to the characteristics of each metal. The advantages are the increase in activity, selectivity and stability and the cost reduction. Metals can be combined in different ways such as in the core-shell bimetallic structure: the cheapest metal forms the core and the most active one (typically a noble metal) constitutes the shell. By adopting this design, the use of rare and expensive metals can be reduced down to 20%.

  One of the future challenges is to find new stable materials, with good activity and especially low cost. Metallic glasses, polymeric carbon nitride (PCN) and materials derived from metal-organic frameworks (MOF) are just a few examples of materials with electrocatalytic properties on which research is currently investing.

Characterization of Nanoparticles

Some techniques that can be used to characterize functionalized nanomaterial catalysts include X-ray photoelectron spectroscopy, transmission electron microscopy, circular dichroism spectroscopy, nuclear magnetic resonance spectroscopy, UV-visible spectroscopy and related experiments.

Homogenous Catalysis

In chemistry, homogeneous catalysis is catalysis in a solution by a soluble catalyst. Homogeneous catalysis refers to catalytic reactions where the catalyst is in the same phase as the reactants. Homogeneous catalysis applies to reactions in the gas phase and even in solids. Heterogeneous catalysis is the alternative to homogeneous catalysis, where the catalysis occurs at the interface of two phases, typically gas-solid. The term is used almost exclusively to describe solutions and often implies catalysis by organometallic compounds.

Homogeneous catalysis using transition metal complexes is an area of research that has grown enormously in recent years. Many remarkable catalytic discoveries have been reported by researchers both in industry and in academia. The area is one of intense research and many practical applications, e.g., the production of acetic acid. Enzymes are examples of homogeneous catalysts.

When materials react to form products, there may be a single reaction or multiple reactions occurring. A single reaction can be called an elementary. For most reactions, the sequence of elementary steps that constitute their mechanism is not known. Sometimes only a single reaction is observed. The reason is that the amount of the intermediate species formed is very small and escapes detection. The speed with which these intermediate species are created and destroyed also makes them difficult to detect. The general scheme is 'reactants --> several intermediate species --> products'. In the case of a non-elementary reaction, the test of the validity of the mechanism proposed is whether the predicted kinetics corresponds to experimental observation. Sometimes reactions may proceed by more than one mechanism, or else their kinetics can be explained by several mechanisms. There are zero-order, first-order, second-order, and higher-order processes.

Advantages

- Homogeneous catalysts are effective at being highly selective towards producing the desired product.

- If a reaction is exothermic, it will release a lot of heat. It is easier to release heat from a solution (as one would do for a homogeneous catalyst) than if one were to use a heterogeneous catalyst, which tends to be an insoluble solid in the solution that adsorbs reaction participants onto it.
- Reactants can easily access the homogeneous catalyst because it is in solution already. This promotes high catalytic activity.
- Species in solution are easier to characterize (e.g. with various spectroscopy techniques) than species adsorbed onto a solid surface.

Disadvantages

- A catalyst capable of dissolving in solution will need to be separated later if it is to be recycled for re-usage (as per the principles of green chemistry).
- The homogeneous catalyst may have issues at high temperatures since it is not desirable for the solution (which contains the catalyst) and any volatile reactants to evaporate, even though high temperatures usually promote faster reactions.

Examples of Homogenous Catalysis

Acid Catalysis

The proton is the most pervasive homogeneous catalyst because water is the most common solvent. Water forms protons by the process of self-ionization of water. In an illustrative case, acids accelerate (catalyze) the hydrolysis of esters:

$$CH_3CO_2CH_3 + H_2O \rightleftharpoons CH_3CO_2H + CH_3OH$$

In the absence of acids, aqueous solutions of most esters do not hydrolyze at practical rates.

Transition Metal-catalysis

Many transformations classically considered homogeneous catalysis utilize soluble coordination and organometallic compounds as catalysts. These catalysts differ from solid catalysts, which are traditionally associated with heterogeneous catalysis. Homogeneous catalysis can be subdivided according to their substrates or their reductive/oxidative character.

Reductions

A prominent class of reductive transformations involve hydrogenation. H_2 is added to unsaturated substrates directly using hydrogen gas or indirectly by transferring hydrogen from substrates. The latter is called transfer hydrogenation. Related reactions entail "HX additions" where X = silyl (hydrosilylation) and CN (hydrocyanation).

Carbonylations

Hydroformylation is a prominent form of carbonylation, involving the addition of H and "C(O)H" across a double bond. MeOH and CO react in the presence of homogeneous catalysts to give acetic acid, as practiced in the Monsanto process and Cativa processes. Carbon monoxide is a common substrate in homogeneous catalysis and diverse reactions have been commercialized, including hydrocarboxylation and hydroesterifications.

Polymerization and Metathesis of Alkenes

A number of polyolefins are produced by Ziegler-Natta catalysis. Alkenes such as ethylene, propylene, and butadiene are common substrates. Olefin metathesis is usually practiced heterogeneously, but homogeneous variants are known.

Oxidations

Homogeneous catalysts are also used in a variety of oxidations. Is the Wacker process, acetaldehyde is produced from ethylene to oxygen. Many non-organometallic complexes are also widely used in catalysis, e.g. for the production of terephthalic acid from xylene. Alkenes are epoxidized and dihydroxylated by metal complexes, as illustrated by the Halcon Process.

Hydration and Hydrolysis

Water is a common reagent in homogeneous catalysis. Esters and amides are slow to hydrolyze in neutral water, but the rates are sharply affected by metal complexes. The fastest and most selective catalysts for such reactions are often metalloenzymes, which can be viewed as large coordination complexes. The hydration of nitriles, alkenes, and alkynes are all catalyzed by metal complexes.

Other Forms of Homogeneous Catalysis

Enzymes are homogeneous catalysts that are essential for life but are also harnessed for industrial processes. A well-studied example is carbonic anhydrase, which catalyzes the release of CO_2 into the lungs from the bloodstream.

Industrial Applications

- Neohexene (3,3-dimethyl-1-butene): An important intermediate in the synthesis of fine chemicals, is produced from the dimer of isobutene, which consists of a mixture of 2,4,4-trimethyl-2-pentene and 2,4,4-trimethyl-1-pentene. Cross-metathesis of the former with ethene yields the desired product. The catalyst is a mixture of WO_3/SiO_2 for metathesis and MgO for isomerization at 370 °C and 30 bar. The isobutene is recycled to the isobutene dimerization unit.

- Vestenamer 8012: A predominantly trans polymer of cyclooctene made with a WCl6 based catalyst (Degussa-Hüls), used in blends and poly-norbornene made by a catalyst generated from $RuCl_3$ in t-BuOH in the presence of air (Norsorex by Atofina, a polymer with a very high molecular weight).
- Polydicyclopentadiene (DCPD): Dicyclopentadiene is the Diels-Alder adduct of cyclopentadiene, an abundant product from the cracker in the refinery. It contains a strained norbornene ring and a less strained cyclopentene ring. ROMP leads to opening of the norbornene ring, but some reaction of the cyclopentene ring also takes place, which leads to cross-linking.

Contrast with Heterogeneous Catalysis

Homogeneous catalysis differs from heterogeneous catalysis in that the catalyst is in a different phase than the reactants. One example of heterogeneous catalysis is the petrochemical alkylation process, where the liquid reactants are immiscible with a solution containing the catalyst. Heterogeneous catalysis offers the advantage that products are readily separated from the catalyst, and heterogeneous catalysts are often more stable and degrade much slower than homogeneous catalysts. However, heterogeneous catalysts are difficult to study, so their reaction mechanisms are often unknown. Enzymes possess properties of both homogeneous and heterogeneous catalysts. As such, they are usually regarded as a third, separate category of catalyst.

Phase-transfer Catalyst

In chemistry, a phase-transfer catalyst or PTC is a catalyst that facilitates the migration of a reactant from one phase into another phase where reaction occurs. Phase-transfer catalysis is a special form of heterogeneous catalysis. Ionic reactants are often soluble in an aqueous phase but insoluble in an organic phase in the absence of the phase-transfer catalyst. The catalyst functions like a detergent for solubilizing the salts into the organic phase. Phase-transfer catalysis refers to the acceleration of the reaction upon the addition of the phase-transfer catalyst.

By using a PTC process, one can achieve faster reactions, obtain higher conversions or yields, make fewer byproducts, eliminate the need for expensive or dangerous solvents that will dissolve all the reactants in one phase, eliminate the need for expensive raw materials and/or minimize waste problems. Phase-transfer catalysts are especially useful in green chemistry—by allowing the use of water, the need for organic solvents is reduced.

Contrary to common perception, PTC is not limited to systems with hydrophilic and hydrophobic reactants. PTC is sometimes employed in liquid/solid and liquid/gas reactions. As the name implies, one or more of the reactants are transported into a second phase which contains both reactants.

Types

Phase-transfer catalysts for anionic reactants are often quaternary ammonium salts. Commercially important catalysts include benzyltriethylammonium chloride, methyltricaprylammonium chloride, methyltributylammonium chloride, and methyltrioctylammonium chloride. Organic phosphonium salts are also used, e.g., hexadecyltributylphosphonium bromide. The phosphonium salts tolerate higher temperatures, but are unstable toward base, degrading to phosphine oxide.

For example, the nucleophilic aliphatic substitution reaction of an aqueous sodium cyanide solution with an ethereal solution of 1-bromooctane does not readily occur. The 1-bromooctane is poorly soluble in the aqueous cyanide solution, and the sodium cyanide does not dissolve well in the ether. Upon the addition of small amounts of hexadecyltributylphosphonium bromide, a rapid reaction ensues to give nonyl nitrile:

$C_8H_{17}Br(org) + NaCN(aq) \rightarrow C_8H_{17}CN(org) + NaBr(aq)$ (catalyzed by a $R_4P^+Br^-$ PTC)

Via the quaternary phosphonium cation, cyanide ions are "ferried" from the aqueous phase into the organic phase.

Subsequent work demonstrated that many such reactions can be performed rapidly at around room temperature using catalysts such as tetra-n-butylammonium bromide and methyltrioctylammonium chloride in benzene/water systems.

An alternative to the use of "quat salts" is to convert alkali metal cations into hydrophobic cations. In the research lab, crown ethers are used for this purpose. Polyethylene glycols are more commonly used in practical applications. These ligands encapsulate alkali metal cations (typically Na^+ and K^+), affording large lipophilic cations. These polyethers have a hydrophilic "interiors" containing the ion and a hydrophobic exterior.

Applications

PTC is widely exploited industrially. Polyester polymers for example are prepared from acid chlorides and bisphenol-A. Phosphothioate-based pesticides are generated by PTC-catalyzed alkylation of phosphothioates. One of the more complex applications of PTC involves asymmetric alkylations, which are catalyzed by chiral quaternary ammonium salts derived from cinchona alkaloids.

Organocatalysis

In organic chemistry, the term organocatalysis (a portmanteau of the terms "organic" and "catalyst") refers to a form of catalysis, whereby the rate of a chemical reaction is increased by an organic catalyst referred to as an "organocatalyst" consisting of carbon, hydrogen, sulfur and other nonmetal elements found in organic compounds. Because of their similarity in composition and description, they are often mistaken as a misnomer for enzymes due to their comparable effects on reaction rates and forms of catalysis involved.

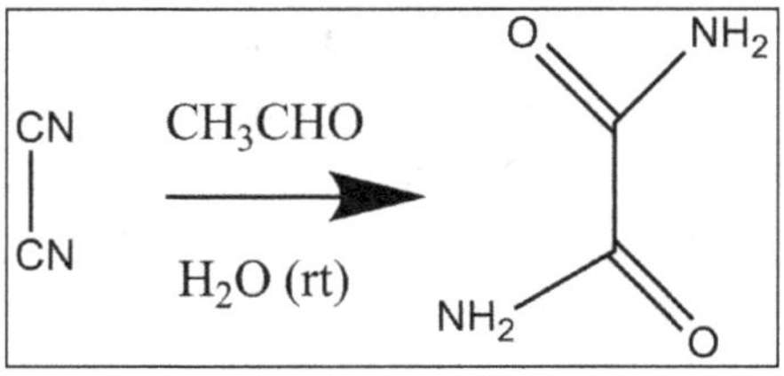

Justus von Liebig's synthesis of oxamide from dicyan and water represents the first organocatalytic reaction, with acetaldehyde further identified as the first discovered pure "organocatalyst", which act similarly to the then-named "ferments", now known as enzymes.

Organocatalysts which display secondary amine functionality can be described as performing either enamine catalysis (by forming catalytic quantities of an active enamine nucleophile) or iminium catalysis (by forming catalytic quantities of an activated iminium electrophile). This mechanism is typical for covalent organocatalysis. Covalent binding of substrate normally requires high catalyst loading (for proline-catalysis typically 20-30 mol%). Noncovalent interactions such as hydrogen-bonding facilitates low catalyst loadings (down to 0.001 mol%).

Organocatalysis offers several advantages. There is no need for metal-based catalysis thus making a contribution to green chemistry. In this context, simple organic acids have been used as catalyst for the modification of cellulose in water on multi-ton scale. When the organocatalyst is chiral an avenue is opened to asymmetric catalysis, for example the use of proline in aldol reactions is an example of chirality and green chemistry.

Regular achiral organocatalysts are based on nitrogen such as piperidine used in the Knoevenagel condensation. DMAP used in esterfications and DABCO used in the Baylis-Hillman reaction. Thiazolium salts are employed in the Stetter reaction. These catalysts and reactions have a long history but current interest in organocatalysis is focused on asymmetric catalysis with chiral catalysts, called asymmetric organocatalysis or enantioselective organocatalysis. A pioneering reaction developed in the 1970s is called the Hajos–Parrish–Eder–Sauer–Wiechert reaction. Between 1968 and 1997, there were only a few reports of the use of small organic molecules as catalysts for asymmetric reactions, but these chemical studies were viewed more as unique chemical reactions than as integral parts of a larger, interconnected field.

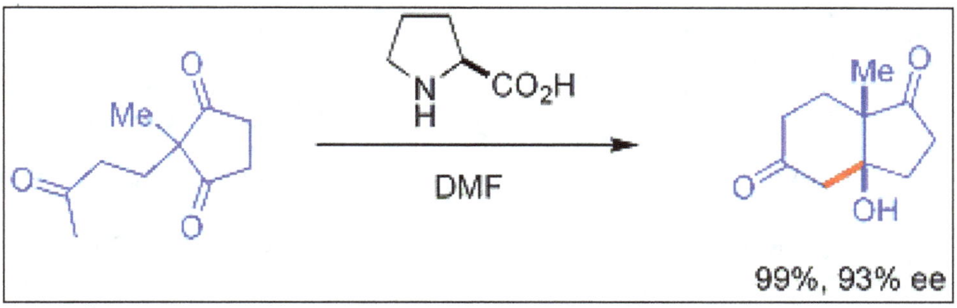

In this reaction, naturally occurring chiral proline is the chiral catalyst in an Aldol reaction. The starting material is an achiral triketone and it requires just 3% of proline to obtain the reaction product, a ketol in 93% enantiomeric excess. This is the first example of an amino acid-catalyzed asymmetric aldol reaction.

The asymmetric synthesis of the Wieland-Miescher ketone is also based on proline and another early application was one of the transformations in the total synthesis of Erythromycin by Robert B. Woodward. Many chiral organocatalysts are an adaptation of chiral ligands (which together with a metal center also catalyze asymmetric reactions) and both concepts overlap to some degree.

Organocatalyst Classes

Organocatalysts for asymmetric synthesis can be grouped in several classes:

- Biomolecules: Proline, phenylalanine. Secondary amines in general. The cinchona alkaloids, certain oligopeptides.
- Synthetic catalysts derived from biomolecules.
- Hydrogen bonding catalysts, including TADDOLS, derivatives of BINOL such as NOBIN, and organocatalysts based on thioureas.
- Triazolium salts as next-generation Stetter reaction catalysts.

Examples of asymmetric reactions involving organocatalysts are:

- Asymmetric Diels-Alder reactions.
- Asymmetric Michael reactions.
- Asymmetric Mannich reactions.
- Shi epoxidation.
- Organocatalytic transfer hydrogenation.

Imidazolidinone Organocatalysis

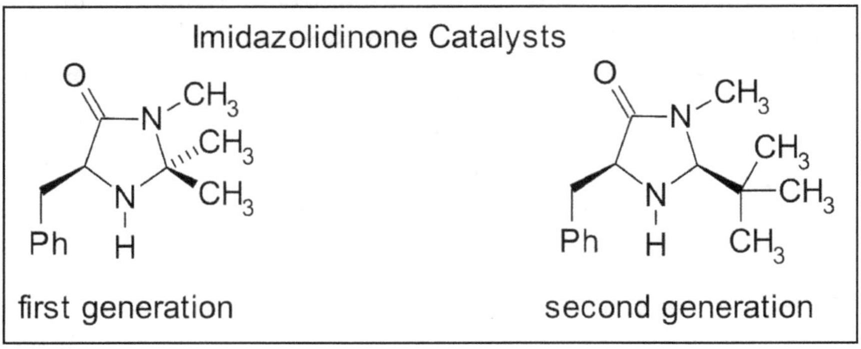

Imidazolidinones are catalysts for many transformations such as asymmetric Diels-Alder reactions and Michael additions. Chiral catalysts induceasymmetric reactions, often with high enantioselectivities. This catalyst works by forming an iminium ion with carbonyl groups of α,β-unsaturated aldehydes (enals) and enones in a rapid chemical equilibrium. This iminium activation is similar to activation of carbonyl groups by a Lewis acid and both catalysts lower the substrate's LUMO:

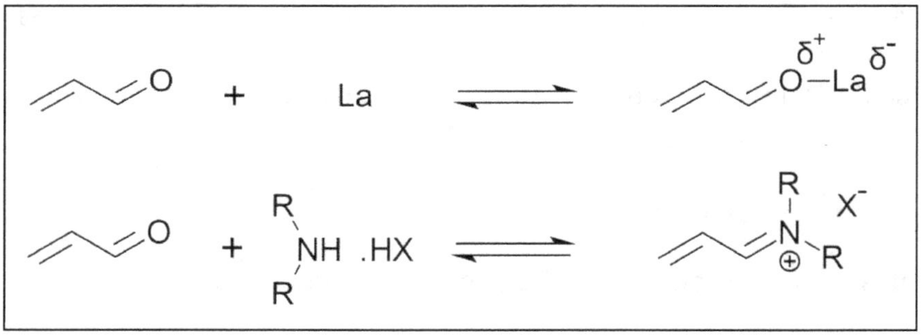

The transient iminium intermediate is chiral which is transferred to the reaction product via chiral induction. The catalysts have been used in Diels-Alder reactions, Michael additions, Friedel-Crafts alkylations, transfer hydrogenations and epoxidations.

One example is the asymmetric synthesis of the drug warfarin (in equilibrium with the hemiketal) in a Michael addition of 4-hydroxycoumarin and benzylideneacetone:

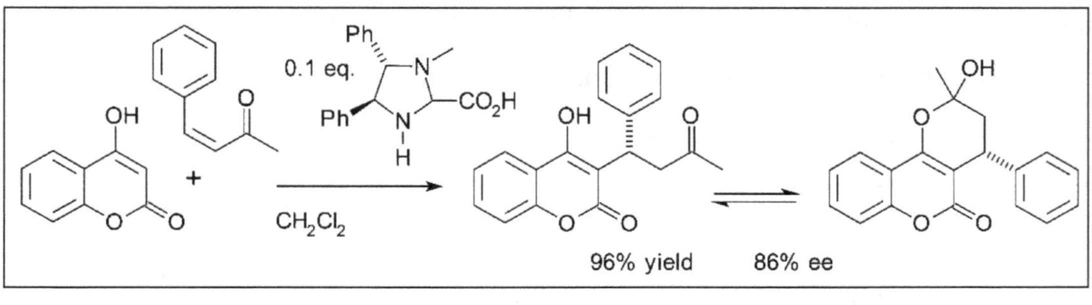

A recent exploit is the vinyl alkylation of crotonaldehyde with an organotrifluoroborate salt:

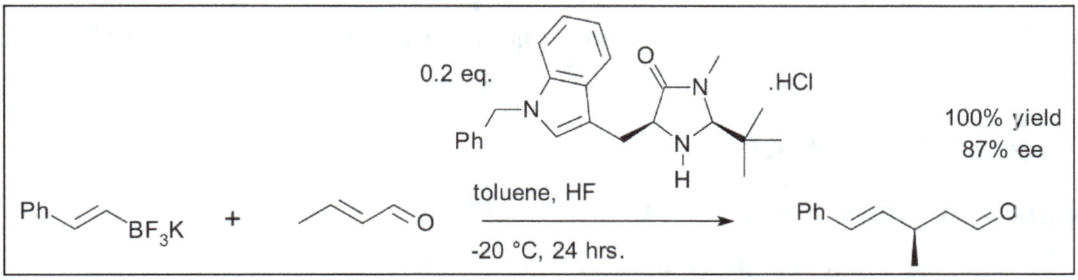

Thiourea Organocatalysis

A large group of organocatalysts incorporate the urea or the thiourea moiety. These catalytically effective (thio)urea derivatives termed (thio)urea organocatalysts provide explicit double hydrogen-bonding interactions to coordinate and activate H-bond accepting substrates.

Photocatalysis

In chemistry, photocatalysis is the acceleration of a photoreaction in the presence of a catalyst. In catalysed photolysis, light is absorbed by an adsorbed substrate. In photogenerated catalysis, the photocatalytic activity (PCA) depends on the ability of the catalyst to create electron–hole pairs, which generate free radicals (e.g. hydroxyl radicals: •OH) able to undergo secondary reactions. Its practical application was made possible by the discovery of water electrolysis by means of titanium dioxide (TiO_2).

Photons from a light source (out of frame on the right hand side) are absorbed by the surface of the titanium dioxide disc, exciting electrons within the material. These then react with the water molecules, splitting it into its constituents of hydrogen and oxygen. In this experiment, chemicals dissolved in the water prevent the formation of oxygen, which would otherwise recombine with the hydrogen.

Types of Photocatalysis

Homogeneous Photocatalysis

In homogeneous photocatalysis, the reactants and the photocatalysts exist in the same phase. The most commonly used homogeneous photocatalysts include ozone and photo-Fenton systems (Fe^+ and Fe^+/H_2O_2). The reactive species is the •OH which is used for different purposes. The mechanism of hydroxyl radical production by ozone can follow two paths.

$$O_3 + h\nu \rightarrow O_2 + O(1D)$$

$$O(1D) + H_2O \rightarrow \bullet OH + \bullet OH$$

$$O(1D) + H_2O \rightarrow H_2O_2$$

$$H_2O_2 + h\nu \rightarrow \bullet OH + \bullet OH$$

Similarly, the Fenton system produces hydroxyl radicals by the following mechanism:

$$Fe^{2+} + H_2O_2 \rightarrow HO\bullet + Fe^{3+} + OH^-$$

$$Fe^{3+} + H_2O_2 \rightarrow Fe^{2+} + HO\bullet 2 + H^+$$

$$Fe^{2+} + HO\bullet \rightarrow Fe^{3+} + OH^-$$

In photo-Fenton type processes, additional sources of OH radicals should be considered: through photolysis of H_2O_2, and through reduction of Fe^{3+} ions under UV light:

$$H_2O_2 + h\nu \rightarrow HO\bullet + HO\bullet$$

$$Fe^{3+} + H_2O + h\nu \rightarrow Fe^{2+} + HO\bullet + H^+$$

The efficiency of Fenton type processes is influenced by several operating parameters like concentration of hydrogen peroxide, pH and intensity of UV. The main advantage of this process is the ability of using sunlight with light sensitivity up to 450 nm, thus avoiding the high costs of UV lamps and electrical energy. These reactions have been proven more efficient than the other photocatalysis but the disadvantages of the process are the low pH values which are required, since iron precipitates at higher pH values and the fact that iron has to be removed after treatment.

Heterogeneous Photocatalysis

Heterogeneous catalysis has the catalyst in a different phase from the reactants.

Heterogeneous photocatalysis is a discipline which includes a large variety of reactions: mild or total oxidations, dehydrogenation, hydrogen transfer, $^{18}O_2$–$^{16}O_2$ and deuterium-alkane isotopic exchange, metal deposition, water detoxification, gaseous pollutant removal, etc.

Most common heterogeneous photocatalysts are transition metal oxides and semiconductors, which have unique characteristics. Unlike the metals which have a continuum of electronic states, semiconductors possess a void energy region where no energy levels are available to promote recombination of an electron and hole produced by photoactivation in the solid. The void region, which extends from the top of the filled valence band to the bottom of the vacant conduction band, is called the band gap. When a photon with energy equal to or greater than the materials band gap is absorbed by the semiconductor, an electron is excited from the valence band to the conduction band, generating a positive hole in the valence band. Such a photogenerated electron-hole pair is termed an exciton. The excited electron and hole can recombine and release the energy gained from the excitation of the electron as heat. Exciton recombination is undesirable and higher levels lead to an inefficient photocatalyst. For this reason efforts to develop functional photocatalysts often emphasize extending exciton lifetime, improving electron-hole separation using diverse approaches that often rely on structural features such as phase hetero-junctions (e.g. anatase-rutile interfaces), noble-metal nanoparticles, silicon nanowires and substitutional cation doping. The ultimate goal of photocatalyst design is to facilitate reactions between the excited electrons with oxidants to produce reduced products, and/or reactions between the generated holes with reductants to produce oxidized products. Due to the generation of positive holes and electrons, oxidation-reduction reactions take place at the surface of semiconductors.

In one mechanism of the oxidative reaction, the positive holes react with the moisture present on the surface and produce a hydroxyl radical. The reaction starts by photo-induced exciton generation in the metal oxide surface (MO stands for metal oxide):

$$MO + hv \rightarrow MO\ (h^+ + e^-)$$

Oxidative reactions due to photocatalytic effect:

$$h^+ + H_2O \rightarrow H^+ + \bullet OH$$

$$2\ h^+ + 2\ H_2O \rightarrow 2\ H^+ + H_2O_2$$

$$H_2O_2 \rightarrow 2\ \bullet OH$$

Reductive reactions due to photocatalytic effect:

$$e^- + O_2 \rightarrow \bullet O_2^-$$

$$\bullet O_2^- + H_2O + H^+ \rightarrow H_2O_2 + O_2$$

$$H_2O_2 \rightarrow 2\ \bullet OH$$

Ultimately, the hydroxyl radicals are generated in both the reactions. These hydroxyl radicals are very oxidative in nature and non selective with redox potential of (E_0 = +3.06 V).

Applications

- Conversion of water to hydrogen gas by photocatalytic water splitting: An efficient photocatalyst in the UV range is based on a sodium tantalite ($NaTaO_3$) doped with La and loaded with a cocatalyst nickel oxide. The surface of the sodium tantalite crystals is grooved with so called nanosteps that is a result of doping with lanthanum (3–15 nm range). The NiO particles which facilitate hydrogen gas evolution are present on the edges, with the oxygen gas evolving from the grooves.
- Use of titanium dioxide in self-cleaning glass: Free radicals generated from TiO_2 oxidize organic matter.
- Disinfection of water by supported titanium dioxide photocatalysts, a form of solar water disinfection (SODIS).
- Use of titanium dioxide in self-sterilizing photocatalytic coatings (for application to food contact surfaces and in other environments where microbial pathogens spread by indirect contact).
- Oxidation of organic contaminants using magnetic particles that are coated with titanium dioxide nanoparticles and agitated using a magnetic field while being exposed to UV light.
- Conversion of carbon dioxide into gaseous hydrocarbons using titanium dioxide in the presence of water. As an efficient absorber in the UV range, titanium dioxide nanoparticles in the anatase and rutile phases are able to generate excitons by promoting electrons across the band gap. The electrons and holes react with the surrounding water vapor to produce hydroxyl radicals and protons. At present, proposed reaction mechanisms usually suggest the creation of a highly reactive carbon radical from carbon monoxide and carbon dioxide which then reacts with the photogenerated protons to ultimately form methane. Although the efficiencies of present titanium dioxide based photocatalysts are low, the incorporation of carbon based nanostructures such as carbon nanotubes and metallic nanoparticles have been shown to enhance the efficiency of these photocatalysts.
- Sterilization of surgical instruments and removal of unwanted fingerprints from sensitive electrical and optical components.
- A less-toxic alternative to tin and copper-based antifouling marine paints, ePaint, generates hydrogen peroxide by photocatalysis.

- Antifouling coatings for filtration membranes, which can also act as separation layer and photocatalyst for degradation of contaminants of emerging concern.
- Decomposition of crude oil with TiO_2 nanoparticles: By using titanium dioxide photocatalysts and UV-A radiation from the sun, the hydrocarbons found in crude oil can be turned into H_2O and CO_2. Higher amounts of oxygen and UV radiation increased the degradation of the model organics. These particles can be placed on floating substrates, making it easier to recover and catalyze the reaction. This is relevant since oil slicks float on top of the ocean and photons from the sun target the surface more than the inner depth of the ocean. By covering floating substrates like woodchips with epoxy adhesives, water logging can be prevented and TiO_2 particles can stick to the substrates. With more research, this method should be applicable to other organics.
- Decontamination of water with photocatalysis and adsorption: The removal and destruction of organic contaminants in groundwater can be addressed through the impregnation of adsorbents with photoactive catalysts. These adsorbents attract contaminating organic atoms/molecules like tetrachloroethylene to them. The photoactive catalysts impregnated inside speed up the degradation of the organics. Adsorbents are placed in packed beds for 18 hours, which would attract and degrade the organic compounds. The spent adsorbents would then be placed in regeneration fluid, essentially taking away all organics still attached by passing hot water counter-current to the flow of water during the adsorption process to speed up the reaction. The regeneration fluid then gets passed through the fixed beds of silica gel photocatalysts to remove and decompose the rest of the organics left. Through the use of fixed bed reactors, the regeneration of adsorbents can help increase the efficiency.
- Decomposition of polyaromatic hydrocarbons (PAHs): Triethylamine (TEA) was utilized to solvate and extract the polyaromatic hydrocarbons (PAHs) found in crude oil. By solvating these PAHs, TEA can attract the PAHs to itself. Once removed, TiO_2 slurries and UV light can photocatalytically degrade the PAHs. The figure shows the high success rate of this experiment. With high yielding of recoveries of 93–99% of these contaminants, this process has become an innovative idea that can be finalized for actual environmental usage. This procedure demonstrates the ability to develop photocatalysts that would be performed at ambient pressure, ambient temperature, and at a cheaper cost.

Nanocatalysts

There has been an explosive growth in the field of green chemistry both in preparing green nanocatalysts as well as green conditions during catalysis of industrially important reactions. Preparing green nanocatalysts refers to synthesizing the nanocatalysts using green solvents or processing the nanocatalysts so that they are finally dispersed in green solvents. Green nanocatalysis refers to doing the catalytic reaction in green solvents and preferably by the use of green nanocatalysts for these reactions. The typical size of green nanocatalysts is 23 nm since this is the optimal size range for many catalytic reactions. Of course, there are also many exceptions to this general trend and other larger sizes of green nanocatalysts have also been synthesized for these purposes. The combination of chiral and nanotechnologies is a promising strategy for green catalytic conversions. The different types of recyclable nanostructured catalytic systems that are used in environmentally friendly organic synthesis. Catalysis with nanoparticles using supercritical solvents with catalyst separation techniques is promising for the development of green chemical processes. Imidazolium-based ionic liquids have been shown to be a good media for the synthesis of metal nanoparticles. These nanocatalysts are efficient green catalysts for several reactions in multiphase conditions. The recent progress in the use of supported metal nanoparticles for sustainable green catalysis is described.

Heterogeneous gold, silver, and copper nanoparticles using hydrotalcite as the support material have been used for environmentally benign organic transformations such as aerobic oxidation of alcohols, lactonization of diols, and deoxygenation of epoxides and nitro aromatic compounds. A variety of new nanocatalysts that have high relevance to green chemistry are discussed such as molecular sieve catalysts for regioselective oxyfunctionalization of alkanes, FischerTropsch catalysts for the production of fuel, enantioselective catalysts, etc .

Synthesis of Green Nanocatalysts

Silica Nanoparticles

Silica nanoparticles that are mild and environmentally benign have been used as catalysts for synthesis of highly substituted pyridines. These catalysts retained most of their catalytic activity after being reused three times. Figure above shows a TEM image of the silica nanoparticles that are used as catalysts for the synthesis of highly substituted pyridines.

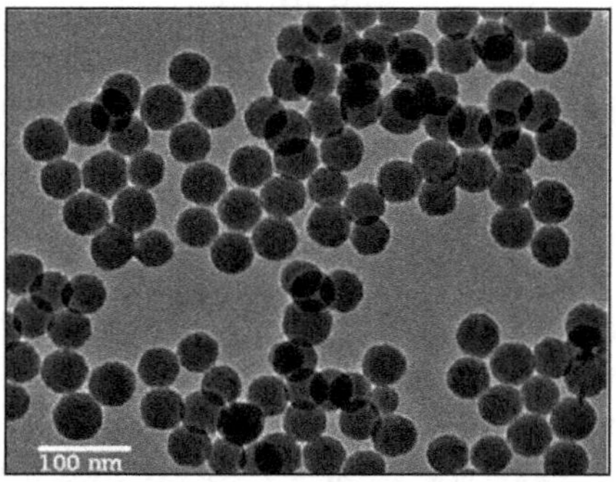

TEM image of silica nanoparticles used as catalysts for the one step, three-component synthesis of highly substituted pyridines.

Silica colloids and nanoparticles can be easily synthesized using the Stoeber synthesis method and redispersed in water which is an environmentally friendly solvent. Environmentally friendly silica nanoparticles have been used as catalysts for the synthesis of thioethers, thioesters, vinylthioethers, and thio-Michael adducts. Very mild, neutral, and reusable silica nanoparticles have been used as catalysts for the synthesis of 4H-pyrans and polysubstituted aniline derivatives. Tubular silica nanostructures are synthesized by environmentally benign and mild conditions. These catalysts are found to be very active and selective in the microwave-assisted homocoupling of terminal alkynes. The three component synthesis of highly substituted pyridines has been conducted using mild, effective, and environmentally benign silica nanoparticles.

Titania Nanoparticles

The titania nanoparticles are used as catalyst using a supercritical carbon dioxide anti-solvent process synthesis method. The size of the titania nanoparticles varies from 27 to 78 nm. Figure shows the SEM and TEM images of the titania nanoparticles.

Palladium Nanoparticles

Palladium nanoparticles can be synthesized using a variety of green methods such as the hydrogen reduction method in water as solvent, ethanol reduction method in an ethanolwater mixture in which the ethanol can be rotovaped, and several other green methods. Palladium nanoparticles capped with polyoxometalate have been used as green electrocatalysts. It was observed that the palladium nanoparticles are highly efficient catalysts for the hydrogen evolution reaction. Palladium nanoparticles that are entrapped in aluminum oxyhydroxide in solvent-free condition have been used as catalysts for the hydrogenation of alkenes and aromatic nitro compounds to form the corresponding alkanes and amino compounds. A green synthesis method has been used to prepare palladium nanoparticles which are used as catalysts for the hydrodechlorination of trichloroethene. The palladium nanoparticles are synthesized by using sodium cellulose as the stabilizer and ascorbic acid as the reducing agent. The Pd nanoparticles have high catalytic activity for the hydrodechlorination reaction of trichloroethene.

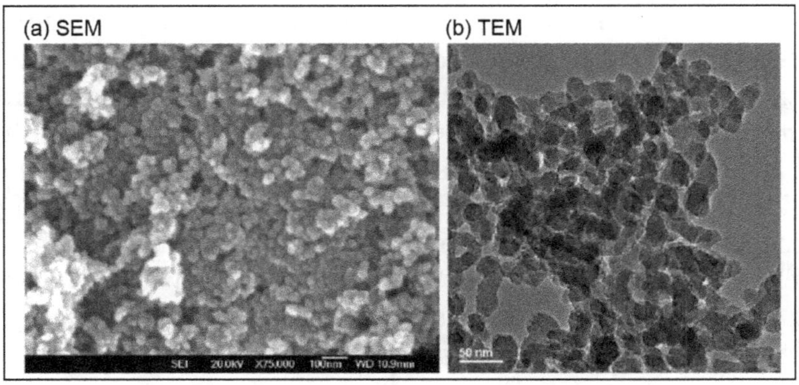

(a) SEM and (b) TEM images of the titania nanoparticle.

The palladium nanoparticles are stabilized using various types of ionic liquids which are green solvents. The palladium nanoparticles have been used as catalysts for the hydrogenation of acetophenone and were observed to be the most efficient catalysts for this reaction. Palladium nanoparticles of different shapes such as nanobelts, nanorods, nanoplates, and nanotrees were synthesized using water as the solvent and vitamin B1 as a reducing agent. These palladium nanoparticles showed excellent catalytic activity for Suzuki, Heck, and Sonogashira reactions. Figure below shows SEM images of the palladium nanoparticles synthesized using vitamin B1 as a reducing agent. A biomimetic method has been used for the synthesis of palladium nanoparticles as model green catalytic systems. These palladium nanoparticles were used as catalysts for the Stille coupling reaction in aqueous solvent. Figure shows the TEM image and size distribution of the palladium nanoparticles that are synthesized using the biomimetic synthesis method.

SEM images of palladium nanoparticles of different shapes that are used as catalysts for the carboncarbon coupling reactions.

Palladium nanoparticles prepared in the presence of ionic liquids have been used as catalysts for the Heck reaction of haloarenes and olefins. Palladiumsepiolite catalysts are synthesized in the presence of ionic liquids. They have been used as catalysts for hydrogenation of alkenes and the Heck cross-coupling reactions. Palladium nanoparticles were used as catalysts for the hydrodechlorination of chloroarenes in water. They have also been used for aerobic oxidation of different alcohols in water to form ketones, aldehydes, and carboxylic acids. Palladium nanoparticles have been synthesized using supercritical carbon dioxide as the solvent and this is a green process since supercritical carbon dioxide is an environmentally benign solvent. Palladium nanoparticles are prepared using a mixture of water and ionic liquid as the solvent and are used to catalyze the Heck cross-coupling reaction. The palladium nanoparticles were found to be very efficient catalysts for the Heck reaction. Figure shows TEM images and electron diffraction data for the palladium nanoparticles prepared in a mixture of water and ionic liquid.

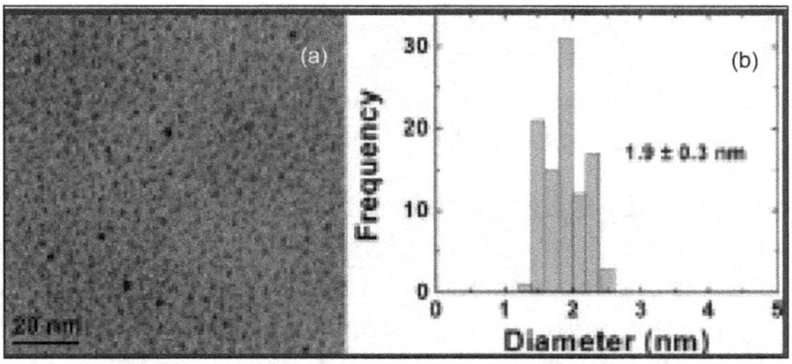

TEM image of the palladium nanoparticles prepared using the biomimetic synthesis method.

Green palladium nanoparticles prepared using CM-cellulose as the stabilizer and ascorbic acid as the reducing agent were used as catalysts for the hydrodechlorination of trichloroethene and were found to have high catalytic activity. Ionic liquids and supercritical carbon dioxide have been used for solventless palladium nanoparticle catalyzed hydrogenations. The nanoparticles are stabilized by the ionic liquids and supercritical carbon dioxide extraction was used for the removal of the metal precursor ligands in the catalyst synthesis step. The catalysts are used for the palladium nanoparticle catalyzed hydrogenation of acetophenone.

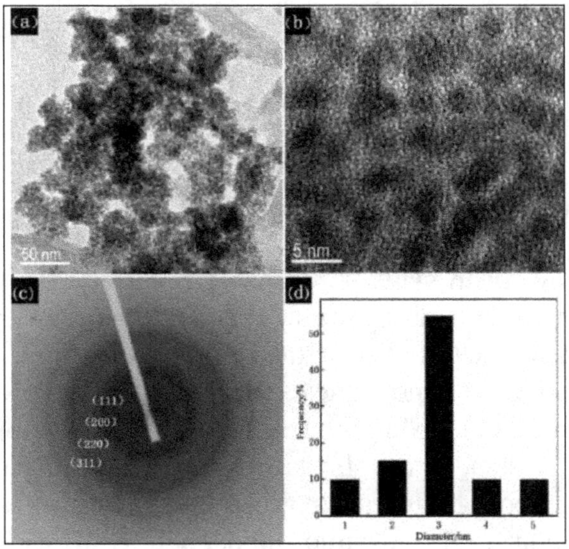

TEM images of the palladium nanoparticles after removing ionic liquid in solution, size distribution histogram and electron diffraction patterns.

Platinum Nanoparticles

Platinum nanoparticles can be synthesized using a variety of green methods such as the hydrogen reduction method in water as solvent, ethanol reduction method in an ethanolwater mixture in which the ethanol can be rotovaped, and several other green methods. Monodisperse green platinum nanoparticles were synthesized by using glucose as the reducing agent and starch as the stabilizer. This synthesis method is simple, environmentally friendly, highly reproducible, and easy to scale up. These nanocatalysts were tested for reduction and oxidation reactions and were found to have high catalytic activity. Monodisperse platinum nanoparticles that are stabilized with ionic liquids have been used as catalysts for four-electron reduction of dioxygen to water.

Figure below shows a TEM image of the platinum nanoparticles that are stabilized using ionic liquids. The environmentally friendly synthesis of platinum nanoparticles decorated carbon nanotubes is reported. The nanohybrid catalyst shows higher electrocatalytic activity for the methanol oxidation reaction compared to traditional platinum-based catalysts. Platinum nanoparticles deposited onto expandable grapheme sheet were synthesized using a green electrochemical route. This catalyst is found to have high catalytic activity for the oxidation of methanol.

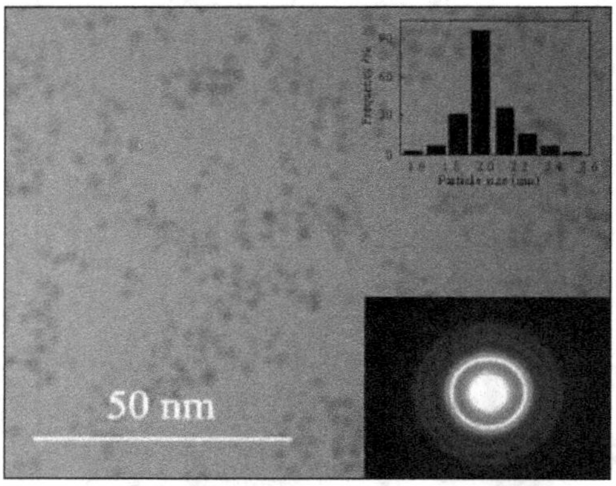

TEM image of platinum nanoparticles that are stabilized in the presence of ionic liquids.

Gold Nanoparticles

Gold nanoparticles have been synthesized using several green methods such as the seed-mediated growth method, doing the synthesis in the presence of ionic liquids, and other reduction methods such as hydrazine reduction method, and sodium borohydride reduction method. Gold nanoparticles supported on mesoporous silica have been used as catalysts for the alkane oxidation reaction. The gold nanoparticles are found to be robust and can be recycled. Figure shows SEM and TEM images of the gold nanoparticles that are formed. Room temperature green imidazolium-based ionic liquids such as 1-butyl-3-methylimidazolium hexafluorophosphate ($BMIMPF_6$) are used as liquid media for the synthesis of gold nanoparticles. It was found that the PVP-capped gold nanoparticles synthesized in the presence of $DMIMPF_6$ are the most stable against aggregation. Gold nanoparticles that are 20nm prepared by addition of $HAuCl_4$ to green tea leaves extract at room temperature. The synthesis of the gold nanoparticles does not involve any toxic chemicals or organic solvents so it is a green synthetic process. The gold nanoparticles are used as catalysts for the reduction of methylene blue dye. Gold nanoparticles have been synthesized by a green photocatalytic method in which the synthesis is conducted in water. Calcium-alginate stabilized gold nanoparticles are prepared using a photochemical green synthetic method. These nanoparticles are used as catalysts for the 4-nitrophenol reduction reaction. Gold/TS-1 nanoparticles are prepared using two green routes which are sol-immobilization method and adsorption reduction method. This gold nanoparticle catalyst was found to have excellent performance for the propylene oxidation reaction.

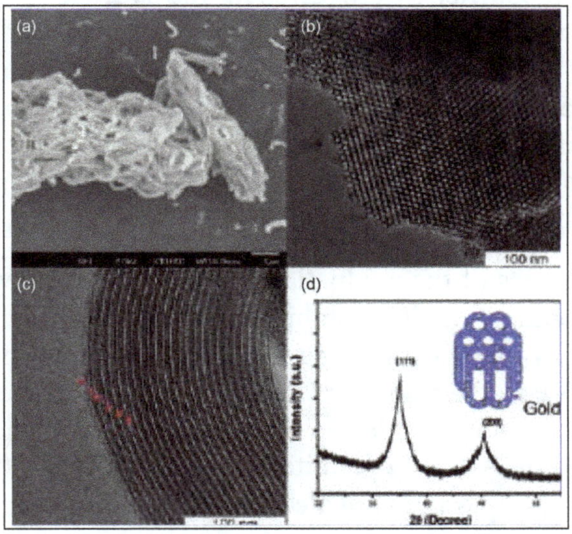

SEM and TEM images of the gold nanoparticles adsorbed onto mesoporous silica.

Silver Nanoparticles

Silver nanoparticles have been synthesized using several green methods such as the seed-mediated growth method, doing the synthesis in the presence of ionic liquids, and other reduction methods such as hydrazine reduction method, and sodium borohydride reduction method. Silver nanoparticles have been synthesized by a green photocatalytic method in which the synthesis is conducted in water. Figure below shows silver nanoparticles that are synthesized using a photocatalytic process. Calciumalginate stabilized silver nanoparticles are prepared using a photochemical green synthetic method. These nanoparticles are used as catalysts for the 4-nitrophenol reduction reaction.

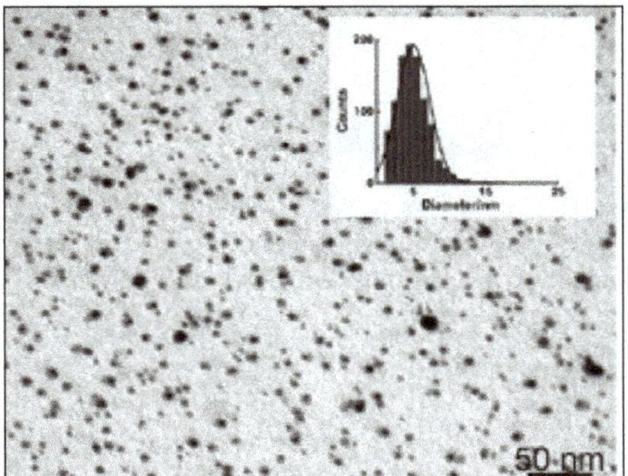

Silver nanoparticles that are synthesized using the green photocatalytic synthesis method.

Palladium-gold Nanoparticles

Room temperature green imidazolium-based ionic liquids such as BMIMPF6 have been used as liquid media for the synthesis of palladium-gold nanoparticles. The catalytic activity of the PdAu nanoparticles were tested for the hydrogenation of 1,3-cyclooctadiene and 3-buten-1-ol and it was observed that the PVP-capped PdAu nanoparticles are the most catalytically active. Gold-palladium nanoparticles that are catalytically active are prepared in the absence of organic ligands and adsorbed onto titanium dioxide supports and is found to be stable in oxidative catalysis conditions. It was found that 70% gold, 30% Pd composition of the bimetallic nanoparticles results in the highest catalytic activity for the oxidative catalysis.

Figure shows TEM images of the goldpalladium nanoparticles which are synthesized without the use of organic liquids as stabilizers. Gold nanoparticles are prepared by using a green synthesis method by using 2,3-dihydroxyfumaric acid (DHFA) as the reducing agent and PVP as the stabilizer. These gold nanoparticles were found to have good catalytic activity for aerial oxidation of hydroxybenzyl alcohols. Ceria is prepared by using supercritical anti-solvent precipitation and serves as a support for goldpalladium nanoparticles for the oxidation of alcohols.

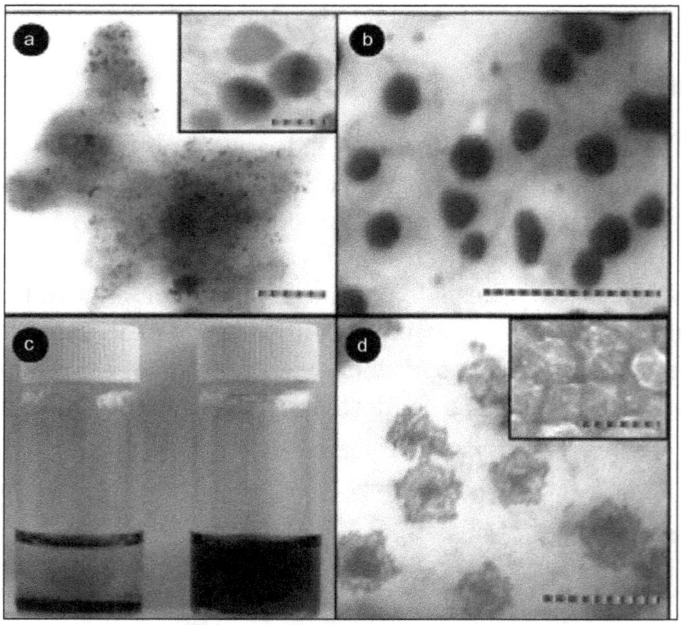

TEM images of the gold-palladium nanoparticle synthesized in the absence of organic liquids as stabilizers.

Nickel-platinum Nanoparticles

Bimetallic nanoparticles have been synthesized using the ethanol reduction method, hydrogen reduction method, and other green methods. Nickel encapsulated by Pt (NiPt) has been synthesized using a green colloidal method. Platinum nanoparticles are very expensive as electrocatalysts so the remedy for this is to reduce the cost by the synthesis of NiPt bimetallic nanoparticles.

Aluminum Oxide Nanoparticles

Solvent-free methods as well as methods involving the use of water as solvent have been used to synthesize aluminum oxide nanoparticles. Aluminum oxide nanoparticles are synthesized in water as the solvent which makes it a green nanocatalyst. It is used as a catalyst for synthesis of 1,5-benzodiazepine and 1,5-benzothiazepine derivatives in mild conditions. The green aluminum oxide nanoparticles are used for the solvent-free synthesis of 2-aryl-2,3-dihydroquinazolin-4(1H)-ones and result in higher yields with shorter reaction times.

Iron Nanoparticles

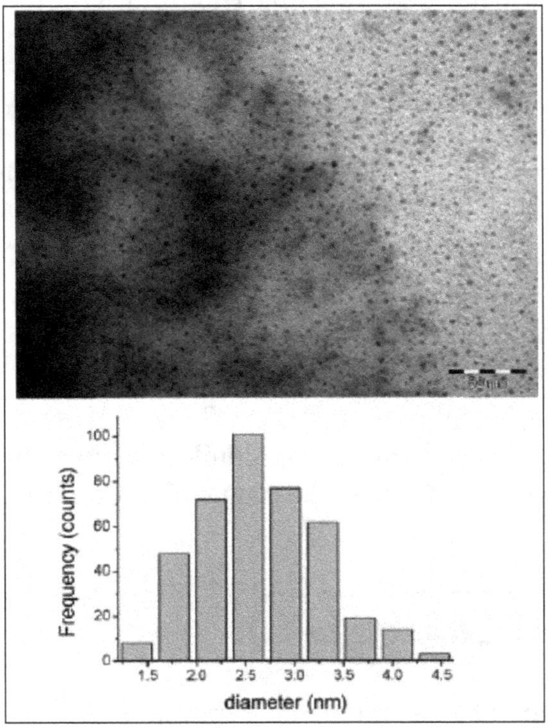

TEM image and size distribution of iron nanoparticles that are used as catalysts for alkene and alkyne hydrogenation reactions

Iron nanoparticles have been synthesized using polymers as capping agents in water as green solvent as well as other types of green methods. A green synthesis of iron nanoparticles has been prepared by using tea polyphenols without the use of additional polymers and surfactants. The iron nanoparticles are used to catalyze the hydrogen peroxide for treatment of organic contamination. Iron nanoparticles have been used as environmentally benign catalysts for alkene and alkyne hydrogenations. Figure below shows the TEM image and size distribution histogram of the iron nanoparticles that has been synthesized as catalysts for environmentally benign alkene and alkyne hydrogenation reactions.

Rhodium Nanoparticles

Rhodium nanoparticles can be synthesized using a variety of green methods such as the hydrogen reduction method in water as solvent, ethanol reduction method in an ethanolwater mixture in which the ethanol can be rotovaped, and several other green methods. Rhodium nanoparticles adsorbed onto titanium dioxide supports are synthesized using water as the solvent. They are also highly active catalysts for the arene hydrogenation reaction in water. The rhodium nanoparticles are stabilized using various types of ionic liquids which are green solvents. The rhodium nanoparticles have been used as catalysts for the hydrogenation of acetophenone. Hydrogenations of different nitrogen-, oxygen-, and sulfur-heterocyclic aromatic compounds have been catalyzed by using rhodium nanoparticles synthesized in water.

Ionic liquids and supercritical carbon dioxide have been used for solventless rhodium nanoparticle catalyzed hydrogenations. The nanoparticles are stabilized by the ionic liquids, and supercritical carbon dioxide extraction was used for the removal of the metal precursor ligands in the catalyst synthesis step. The catalysts are used for the rhodium nanoparticle catalyzed hydrogenation of acetophenone.

Zinc Oxide Nanoparticles

A green synthesis of ZnO nanoparticles has been developed in which a new leucine-based diamine amphiphile was synthesized and self-assembled. In the presence of Zn^{2+} ions, the leucine-based diamine amphiphile assembled into nanofibers that efficiently formed ZnO nanoparticles upon heating in the presence of $Zn(CH_3COO)_2$.

Bifunctional Catalysts

Bifunctional catalysts are able, as the name implies, to catalyse the conversion of one compound to another, using two substances on the surface. For example, in reforming naphtha (a mixture of straight chain alkanes, with 6-10 carbon atoms) a bifunctional catalyst is used. The most well known one is platinum impregnated on the surface of alumina and both the metal and the oxide play their parts in the process. The first step is the dehydrogenation of the alkanes to alkenes, catalysed by the metal, followed eventually by adsorption of the alkene molecules on alumina. Because platinum is involved, the reforming is sometimes called platforming. The hydrogen ensures that the resulting alkenes and cycloalkenes subsequently react with hydrogen to form saturated compounds. In this example butane is dehydrogenated to butene.

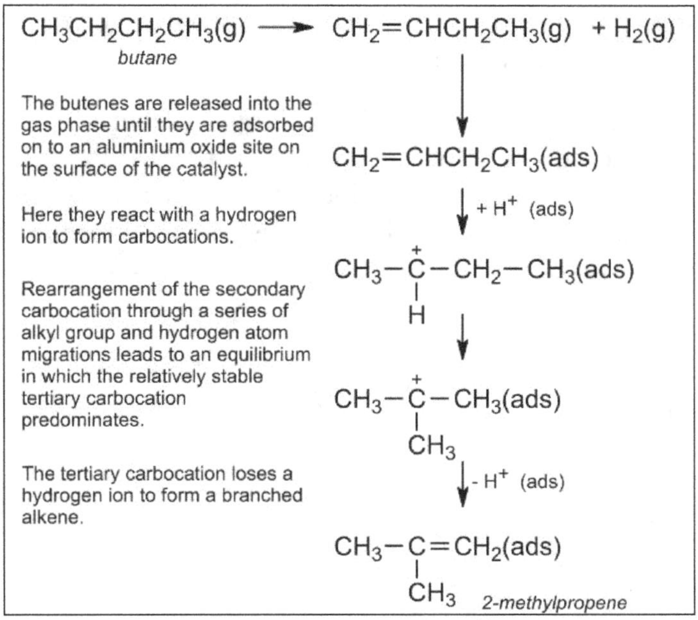

A mechanism for the reforming of butane to 2-methylpropene (isobutene).

The branched alkene molecule is desorbed into the gas phase until it is readsorbed on to a metal site where it is hydrogenated to form a branched alkane, 2-methylpropane (isobutane). In the industrial process, naphtha vapour is passed over platinum and rhenium (ca 0.3% each) which are finely dispersed over aluminium oxide.

The rhenium is thought to play an interesting role. If a sulfur compound is allowed to pass over the surface of the catalyst, it is preferentially adsorbed by the rhenium. If sulfur compounds are not removed, reactions occur leading eventually to the formation of carbon which causes the activity of the catalyst to be markedly reduced.

Branched alkanes have a much higher octane rating than straight chain ones. Not only are the alkanes now branched, but cycloalkanes are also formed and, from them, aromatic hydrocarbons. All three classes of hydrocarbon have a higher octane rating than naphtha. Besides aluminium oxide and silicon dioxide, other oxides are important catalysts. For example, in the Contact Process used to manufacture sulfuric acid, the catalyst for the oxidation of sulfur dioxide to sulfur trioxide is vanadium(V) oxide on the surface of silica:

$$2SO_2(g) + O_2(g) \rightarrow 2SO_3(g)$$

Potassium sulfate is added as a promoter. Its mode of action is not absolutely clear but it appears to be because its presence lowers the melting point of the catalyst, and allows it to spread out as a very thin layer over the entire surface.

Several important industrial processes are catalysed by mixed metal oxides. The surfaces contain two or more different metal atoms, O^{2-} ions and -OH groups. They are quired. For example, propene can be oxidized to propenal (acrolein) using a mixture of bismuth(III) and molybdenum(VI) oxides.

Without the catalyst, propene is oxidized to a large number of organic compounds, including methanal and ethanal, and eventually forming carbon dioxide. The oxygen atoms on the surface of molydenum(VI) oxide are not very reactive, reacting selectively with propene and breaking the weakest bond in the alkene to form an allyl radical:

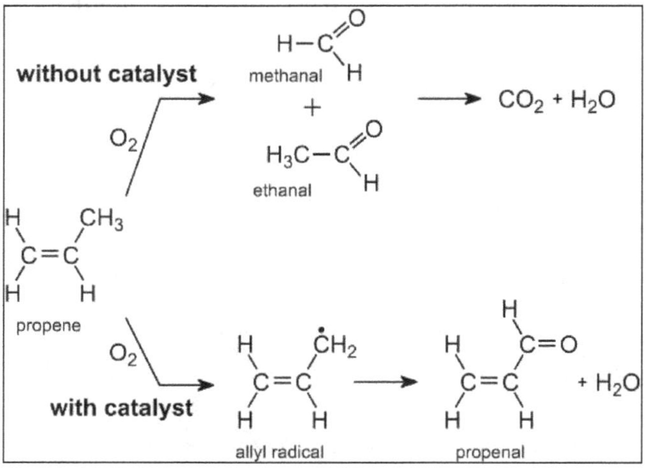

The oxidation of propene.

The allyl radical is then oxidized on the surface to yield propenal. It is postulated that the allyl radical is oxidized by an oxygen atom that is adsorbed at a molybdenum site. Another oxygen atom, adsorbed on a bismuth site, is then transported to the reduced molybdenum site to replace that oxygen. There is a compensating transport of electrons to complete the cycle.

The same catalyst is also used to manufacture propenonitrile:

$$\underset{\text{propene}}{CH_2=CH-CH_3}(g)+1\tfrac{1}{2}O_2(g)\rightarrow \underset{\text{propenonitrile}}{CH_2=CH-CN}(g)+3H_2O(g)$$

References

- Liese, Andreas; Seelbach, Karsten; Wandrey, Christian, eds. (2006). Industrial Biotransformations (2nd ed.). John Wiley & Sons. p. 556. ISBN 978-3527310012
- Introduction-to-catalysis, background-to-green-chemistry: learning.chem21.eu, Retrieved 28 July, 2020
- Briggs GE, Haldane JB (1925). "A Note on the Kinetics of Enzyme Action". The Biochemical Journal. 19 (2): 338–9. doi:10.1042/bj0190338. PMC 1259181. PMID 16743508
- Marigo, Mauro; Kjærsgaard, Anne; Juhl, Karsten; Gathergood, Nicholas; Jørgensen, Karl Anker (23 May 2003). "Direct Catalytic Asymmetric Mannich Reactions of Malonates and -Keto Esters". Chemistry: A European Journal. 9 (10): 2359–2367

Green Reagent and Green Solvent

5

- **Dimethyl Carbonate**
- **Polymer Supported Reagents**
- **Green Solvents**

Green reagent deals with the reduction or elimination of the use or generation of hazardous substances in the design, manufacture and application of chemical products. The biosolvents that are derived from the processed agricultural crops are called green solvent. All the aspects related to green reagent and solvent have been carefully analyzed in this chapter.

Green reagent is one that reduces or eliminates the use or generation of hazardous substances in the design, manufacture and application of chemical products. Chemistry is undeniably a very prominent part of our daily lives. Chemical developments also bring new environmental problems and harmful unexpected side effects, as a result there is a need for 'greener' chemical products. A famous example for environmental pollutant is the pesticide DDT. Green reagents have minimal side effects, and low pollutability.

In order to carry out the transfonnation of selected feedstock into the target molecule the criteria of efficiency, availability and effect of the reagent used must be kept in mind.

Dimethyl Carbonate

Dimethyl carbonate stands out as one of the key chemicals in the Green Chemistry approach thanks to its versatile uses as a safe, non-toxic, green reagent. Among its uses, its capability of replacing dimethyl sulfates and methyl halides in methylation reactions, as well as replacing phosgene in the production of polycarbonates, must be highlighted. It is also being considered as a component of reformulated fuels due to its high oxygen content, having as much oxygen as three times the amount in methyl-tert-butyl ether (MTBE), and good blending properties. Its molecule is shown in figure.

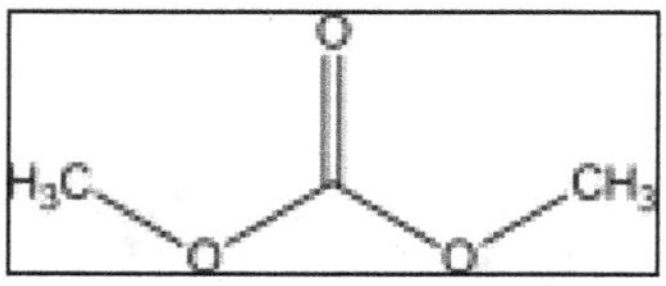

Dimethyl carbonate molecule.

Despite being able to be synthesized by numerous routes, DMC is only industrially produced by three different routes: liquid phase oxidative carbonylation of methanol, vapor phase oxidative carbonylation of methanol and transesterification of ethylene carbonate with methanol.

Traditionally, DMC was prepared by phosgenation of methanol, but that route is being gradually abandoned due to the environmental, safety and economic drawbacks that it entails.

Focusing on the objective of the paper, the urea route allows the production of DMC using only urea and methanol as raw materials, which are comparatively abundant and have a low price. It also avoids the use of phosgene, carbon monoxide, nitric oxide and ethylene oxide, which are present in the other routes. In addition, it also provides the perfect opportunity for the integration of both the fuel industry and the fertilizer industry, since urea production is comparable to the environmental gasoline oxygenates demand.

As such, this route seems very attractive for the industry. It has however some drawbacks in which this paper dwells on.

Carbon Dioxide Utilization for Global Sustainability

Dimethyl carbonate (DMC) has attracted much attention in terms of a non-toxic substitute for dimethyl sulfate and phosgene, which are toxic and corrosive methylating agents. In addition, DMC is considered to be an option for meeting the oxygenate specifications for the transportation fuel . The traditional synthesis of DMC used to require phosgene as a reagent. There are two processes based on the oxy-carbonylation of methanol have been already successfully developed up to larger scale: (1) the oxidative carbonylation of CH_3OH with carbon monoxide and oxygen catalyzed by cuprous chloride, and (2) an oxidative carbonylation process using a palladium catalyst and methyl nitrite promoter.

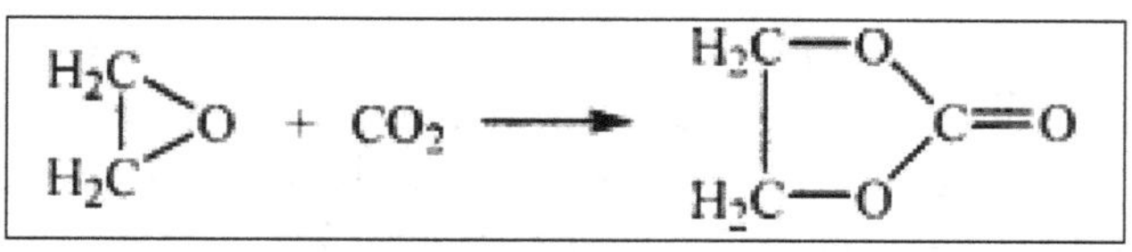

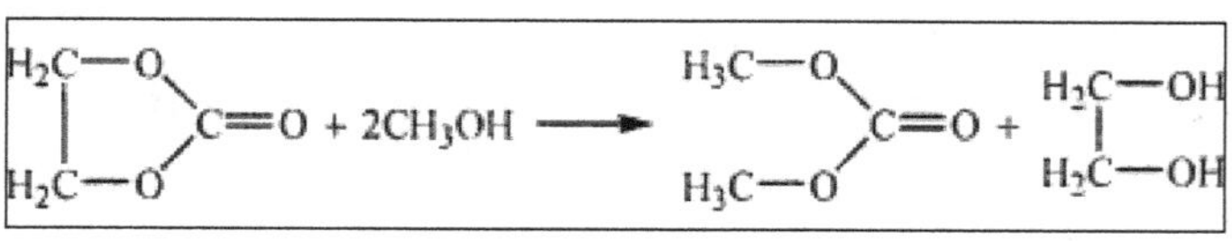

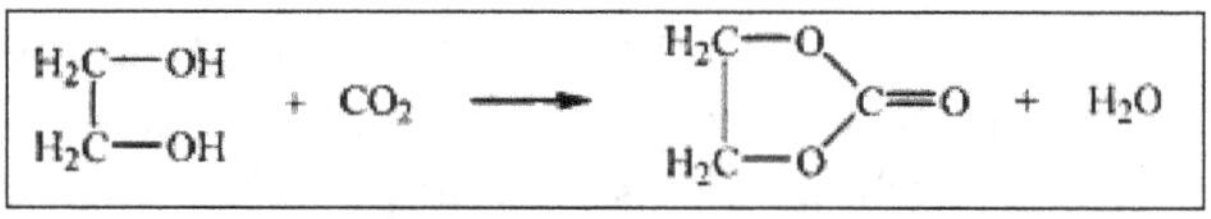

The utilization of carbon dioxide as the raw material of the DMC synthesis process has been attempted. One possible reaction scheme is ethylene carbonate formation from ethylene oxide and CO_2, and the transesterification of ethylene carbonate (EC) with methanol.In this system, ethylene glycol (EG) is always co-produced with DMC. If ethylene carbonate can be synthesized from EG and CO_2, the reaction scheme becomes more valuable since EG can be reused. However, the synthesis of EC from EG and CO_2 has not been reported. Some catalysts have been reported to be effective for the synthesis of DMC from CH_3OH and CO_2. Recently we have found that ZrO_2 is effective to selective DMC formation from CH_3OH and CO_2. In addition, it is found as a result of catalyst development that CeO_2-ZrO_2 solid solution catalysts exhibited much higher activity than ZrO_2, especially at lower reaction temperature.

Production

Production of dimethyl carbonate worldwide is limited to Asia, the Middle East, and Europe.

Dimethyl carbonate is traditionally prepared by the reaction of phosgene and methanol. Methyl chloroformate is produced as an intermediate:

$$COCl_2 + CH_3OH \rightarrow CH_3OCOCl + HCl$$

$$CH_3OCOCl + CH_3OH \rightarrow CH_3OCO_2CH_3 + HCl$$

This synthesis route has been largely replaced by oxidative carbonylation. In this process, carbon monoxide and an oxidizer provide the equivalent of CO^{2+}.

$$CO + 1/2\ O_2 + 2\ CH_3OH \rightarrow (CH_3O)_2CO + H_2O$$

It can also produced industrially by a transesterification of ethylene carbonate or propylene carbonate and methanol, which also affords respectively ethylene glycol or propylene glycol.

Reactions and Potential Applications

Methylating Agent

Dimethyl carbonate methylates anilines, phenols, and carboxylic acids, but many of these reactions require the use of an autoclave.

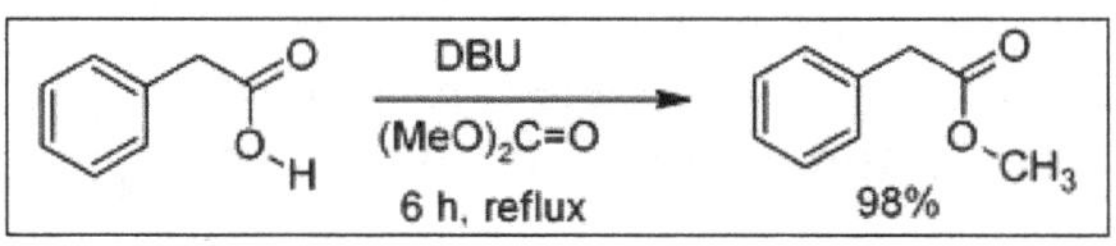

Dimethyl carbonate's main benefit over other methylating reagents such as iodomethane and dimethyl sulfate is its low toxicity. Additionally, it is biodegradable. Unfortunately, it is a relatively weak methylating agent compared to these traditional reagents.

Solvent

In the US, dimethyl carbonate was exempted under the definition of volatile organic compounds (VOCs) by the U.S. EPA in 2009. Due to its classification as VOC exempt, dimethyl carbonate has grown in popularity and applications as a replacement for methyl ethyl ketone (MEK), tert-butyl acetate, and parachlorobenzotrifluoride. Dimethyl carbonate has an ester- or alcohol-like odor, which is more favorable to users than most hydrocarbon solvents it replaces. Dimethyl carbonate has an evaporation rate of 3.22 (butyl acetate = 1.0), which slightly slower than MEK (3.8) and ethyl acetate (4.1), and faster than toluene (2.0) and isopropanol (1.7). Dimethyl carbonate has solubility profile similar to common glycol ethers, meaning dimethyl carbonate can dissolve most common coating resins except perhaps rubber based resins. Hildebrand solubility parameter is 20.3 MPa and Hansen solubility parameters are: dispersion = 15.5, polar = 3.9, H bonding = 9.7. Dimethyl carbonate is partially soluble in water up to 13%, however it is hydrolyzed in water-based systems over time to methanol and CO_2 unless properly buffered. Dimethyl carbonate can freeze at same temperatures as water, it can be thawed out with no loss of properties to itself or coatings based on dimethyl carbonate.

Intermediate in Polycarbonate Synthesis

A large captive use of dimethyl carbonate is for the production of diphenyl carbonate through transesterification with phenol. Diphenyl carbonate is a widely used raw material for the synthesis of bisphenol-A-polycarbonate in a melt polycondensation process. The most common route is described by Fukuoka, of the Japanese "Asahi Kasei" chemical corporation.

Alternative Fuel Additive

There is also interest in using this compound as a fuel oxygenate additive.

Polymer Supported Reagents

The polymer supported reagents offer simplified reaction procedure, which eliminates the tedious chromatographic separation methods, which are usually employed in organic reactions. This enables the use of excess polymeric reagent to obtain high yields of the product. Regeneration and reusability are some other important advantages of the polymeric reagents.

There are a group of reagents which though are ordinary reagents, are bound to polymer support. The main advantage of using these reagents is that any excess of the reagent can be recovered by filtration and used again. Also, the isolation of the product is very easy. Some of such reagents are given as follows:

Polymer Supported Peracids

These are used for epoxidations of alkenes in good yields:

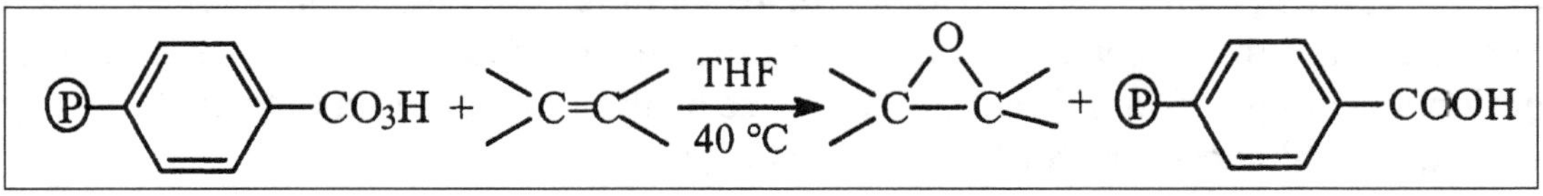

The polymer supported chromic acid (Amberlyst A-26, $HCrO_4^-$ form is commercially available) and has been used to oxidise primary and secondary alcohol to carbonyl compounds6 and also oxidizes allylic and benzylic halides to aldehydes and ketones.

Polystyrene methyl sulphide on reaction with chlorine in the presence of triethylamine forms s-chloro sulfonium chloride resin, which acts as a selective oxidant for the alcohols.

It is an efficient polymer based brominating agent and is used as benzylic and allylic brominating agent. Thus, cumene on bromination9 yield α,β,β'– tribromocumene. However, bromination with NBS gives a α – ,dibromocumene and a-bromocumene.

The use of polymer-supported reagents and scavengers provides an attractive and practical method for the clean and efficient preparation of novel chemical entities. These methods can be extended in a multistep fashion to provide access to more complex structures, including biologically active natural products. supported reagents, catalysts and scavenging agents as useful directory to assist with future synthesis planning in the chemical community. group has a long track record in the development and application of polymer-supported reagents in organic synthesis

Development of Polymer Supported Reagents

These are a small selection of specifically developed reagents to tackle particular synthesis challenges.

Polymer Supported Iridium Catalyst

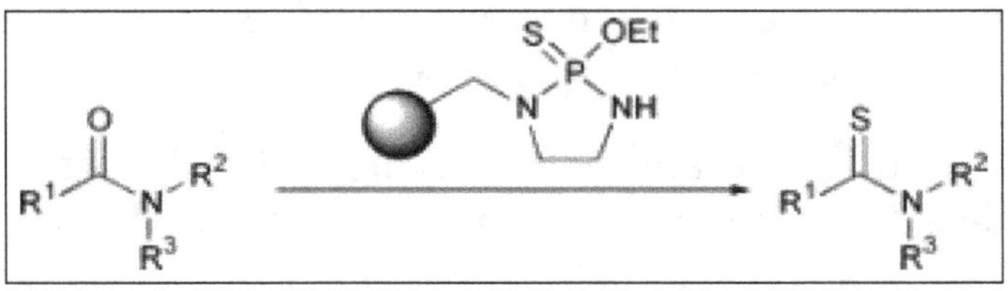

We prepared a polymer-supported iridium catalyst and used in the isomerisation of the double bonds in aryl allylic derivatives with excellent trans selectivity and without the need for conventional work-up procedures.

Polymer Supported Thiolating Agent

We developed a thiolating reagent for the conversion of carbonyls to thiocarbonyls and demonstrated its use on a range of amides. Secondary or tertiary amides were converted cleanly and efficiently through to the corresponding thioamides and primary amides were converted to the corresponding nitriles. While reactions could be aided by conventional heating, we found that if microwave heating was used, in the presence of an ionic liquid, enhanced reaction rates are achieved.

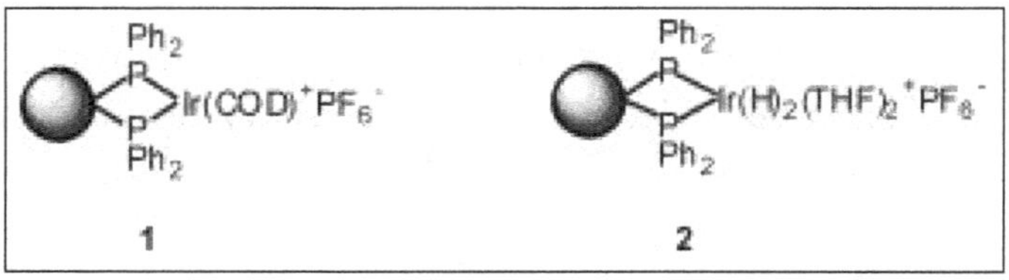

Polymer Supported Perruthenate

Prepare a polymer supported perruthenate reagent. It is used in the conversion of primary and secondary alcohols to aldehydes and ketones, respectively, which afforded pure products without the need for conventional work-up procedures.

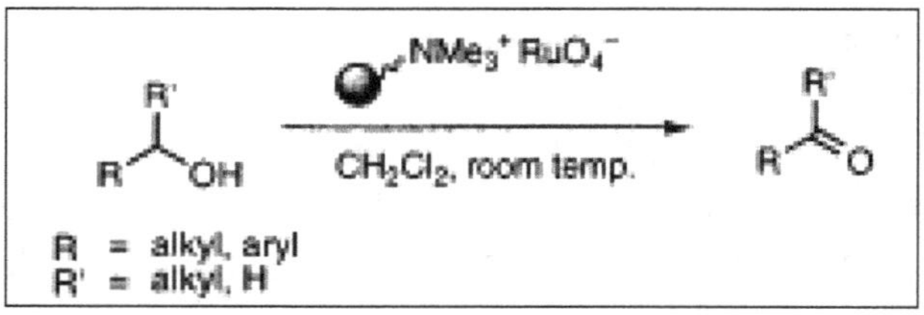

The Polymer Supported Reagent Approach to Natural Product Synthesis

Oxomaritidine and Epimaritidine

Our first serious application of supported reagents for natural product synthesis was published in 1999. We reported concise routes to two amaryllidaceae alkaloids; oxomaritidine and epimaritidine in just five and six steps, respectively. Supported reagents, featured in all of the steps, were used in a sequential fashion and led to pure products following a simple filtration to remove spent reagents.

The route began with the quantitative oxidation of 3,4-dimethoxybenzyl alcohol to the corresponding aldehyde 4 using our own polymer-supported perruthenate (PSP), a catalytic oxidant. Next, the reductive amination of aldehyde 4 was achieved firstly by coupling with the phenolic amine 5 and followed by reduction using a polymer-supported borohydride reagent to furnish the corresponding secondary amine 6 under previously optimised conditions. Amine 6 was protected as trifluoroacetate 7 by treatment with trifluoroacetic anhydride and immobilized dimethylaminopyridine (PS-DMAP) as the catalyst and base. The resulting product 7 underwent smooth oxidative coupling to the spirodienone 8 using polymer-supported hypervalent iodine diacetate, a reagent developed specifically for the task. In order to guarantee good conversion in this process, it was essential to use trifluoroethanol as the solvent. Finally, a wet carbonate ion-exchange resin simultaneously deprotected and initiated cyclisation (via conjugate addition) to give the first of the natural products, oxomaritidine, in essentially quantitative yield after filtration and solvent evaporation. Subsequent stereoselective reduction of oxomaritidine, using an immobilized copper boride (or nickel boride) equivalent, delivered the second natural product epimaritidine in an excellent 50% overall yield over the six step sequence. The synthesis was scaleable to deliver gram quantities.

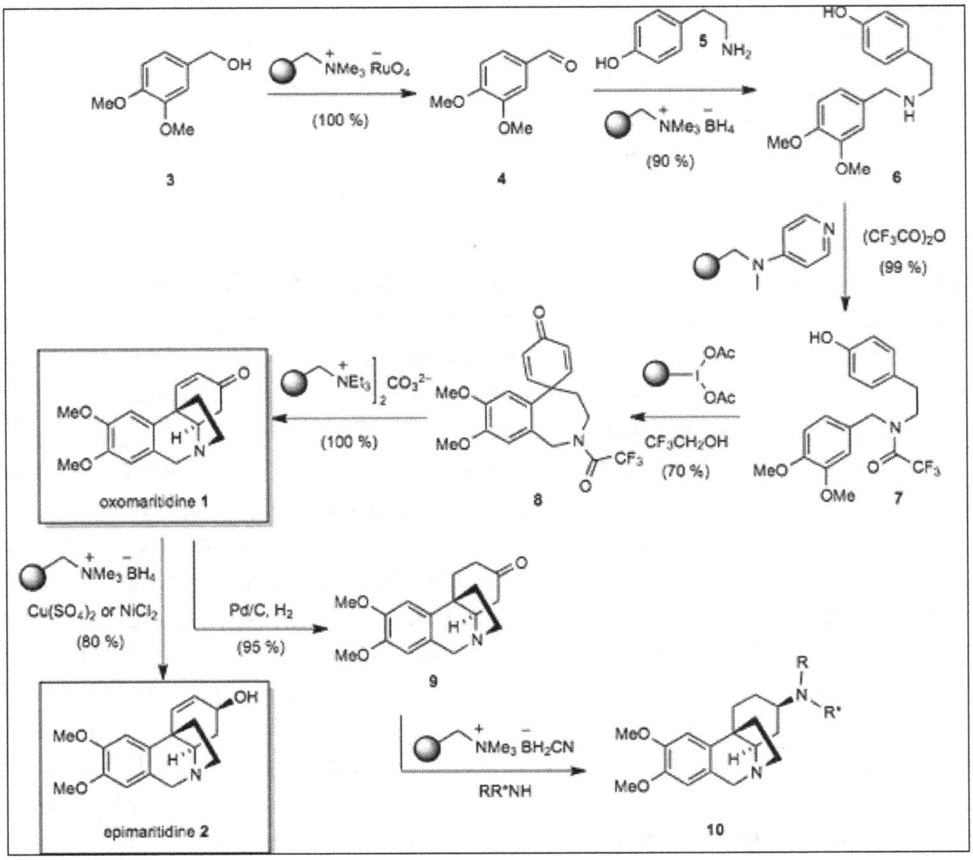

Epibatidine

The potent analgesic alkaloid epibatidine isolated from the Equadorian poison frog Epipedobates tricolour, was a more challenging compound to synthesize using immobilization techniques. This synthesis involved the orchestrated employment of 10 polymer-supported reagents and scavengers to give epibatidine in greater than 90% purity without any chromatographic steps.

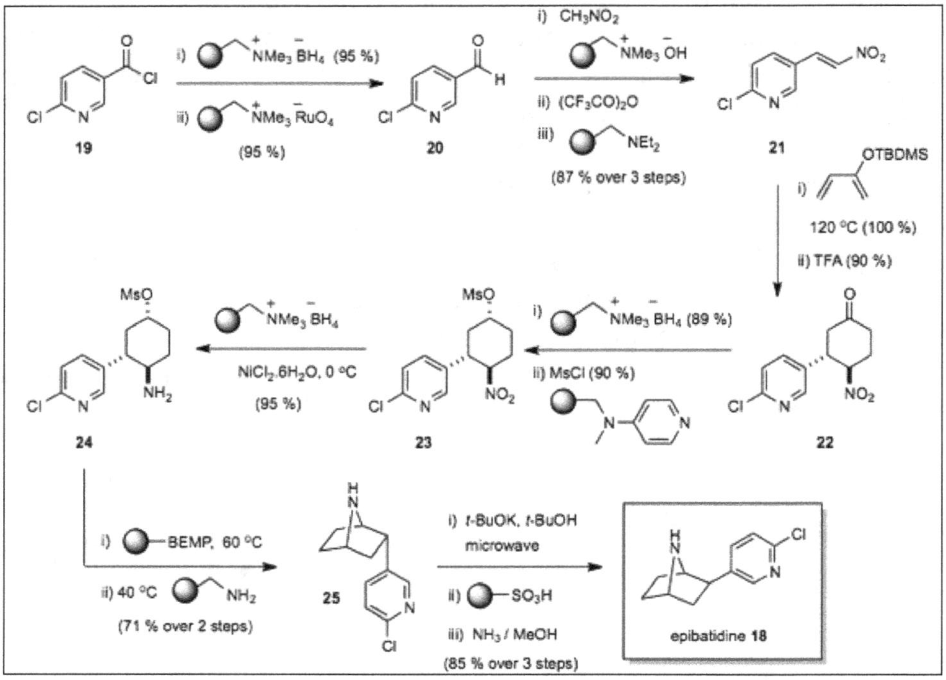

The synthesis began by transforming the commercially available acid chloride 19 to aldehyde 20. This was achieved in a two-step process by reduction of the acid chloride to the intermediate alcohol with polymer-supported borohydride, then partial re-oxidation by the PSP reagent to deliver aldehyde 20. Alternative immobilized oxidants such as supported permanganate and diacetoxyiodobenzene were equally efficient. Oxidants such as Magtrieve (magnetised CrO_2 and MnO_2), although performing the reaction, were less suitable as they required considerably longer reaction times. Aldehyde 20 then underwent straightforward conversion to the nitrostyrene 21. A Henry reaction promoted by the basic Amberlite resin (IRA 420 OH^- form) and followed by elimination of an intermediate trifluoroacetate using polymer-supported diethylamine as a base gave 21. By NMR analysis at each stage, products were shown to be better than 95% pure. In an important modification to the above sequence of reactions, it was found that by the incorporation of supported reagents contained in sealed porous polymer pouches the conversion of chloride 19 to nitrostyrene 21 was possible in a one-pot operation. Thus, when an individual reaction was deemed to be complete the pouch was simply removed, washed with solvent, and the next reagent pouch added to the flask. This process obviated the need for filtration between individual steps.

In the next phase of the synthesis, a regioselective Diels-Alder reaction of nitrostyrene 21 with 2-tert-butyldimethylsilyloxybutadiene in a sealed tube at 120 °C and work-up with a volatile acid (TFA) to hydrolyse the intermediate enol ether gave ketone 22 exclusively as the trans-substituted product. Stereoselective reduction of the carbonyl and corresponding mesylate formation gave 23, again using an immobilized suite of reagents. Selective reduction of the nitro 23 to amine 24 in the presence of other sensitive functionalities was next achieved using polymer-supported nickel boride. Treatment of amine 24 with PS-BEMP initiated an intramolecular displacement of the mesylate which, following a scavenging step with a polystyrene aminomethyl resin to remove excess mesylate, yielded the natural product precursor 25. Epimerisation to epibatidine was readily achieved by microwave heating in the presence of potassium tert-butoxide. The natural product was ultimately isolated by a catch-and-release technique.

Plicamine

Much of the above constitutes the preparation of relatively simple structures. However, these syntheses actually helped prepare the ground for more challenging targets such as the synthesis of the alkaloid (+)-plicamine. Extensive use of parallel optimisation methods and focussed microwave techniques, to achieve fast reaction times, were used to complete the total synthesis in just six weeks without rehearsal of any reactions using conventional solution-phase methods or separation techniques. In this respect this synthesis stands out. How the route (in multi-gram quantities) can be modified to afford analogues or generate related structural scaffolds for further chemical decoration. Also of note is that since this synthesis relied on a single asymmetric centre in the starting material to control all the others, by use of the opposite enantiomer, the unnatural (−)-plicamine enantiomer can also be prepared and examined for biological activity.

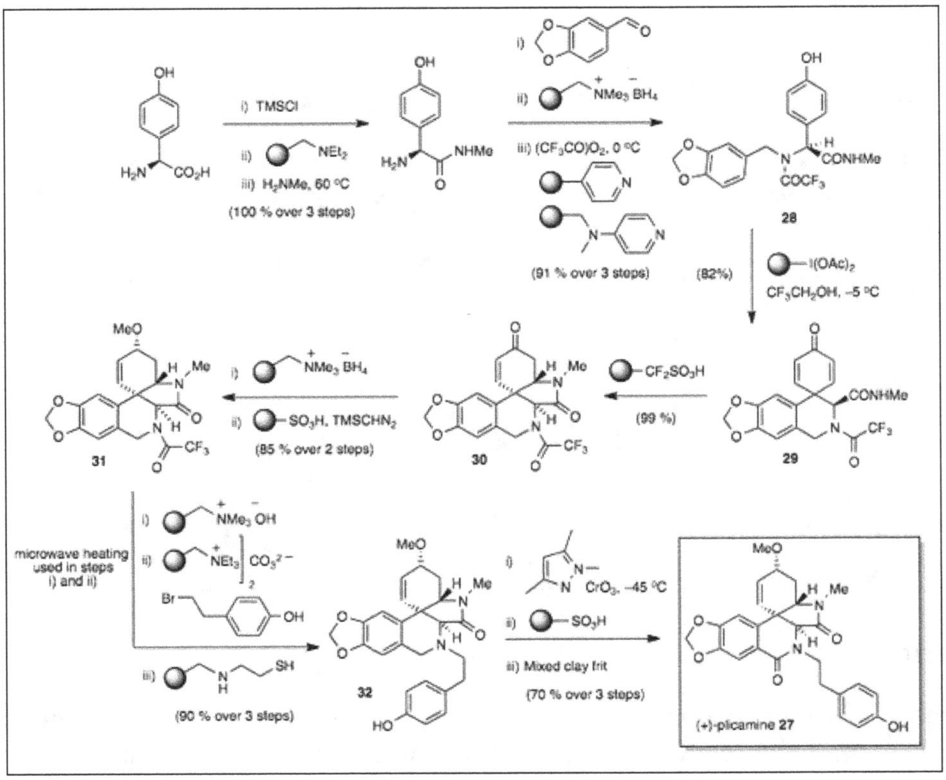

Polymer-supported hypervalent iodine reagent performed well to convert phenol 28 to spirodienone 29. Nafion-H (fluorosulfonic acid resin) catalysed the final cyclisation of 29 to form the tricyclic core of the natural product in virtually quantitative yield. After stereo- and regioselective reduction of 30 using supported borohydride, the very hindered intermediate alcohol was methylated by treatment with trimethylsilyl diazomethane and supported sulfonic acid resin to give 31. This process is very mild and is to be recommended in difficult situations. The remaining steps to (+)-plicamine from 31 were relatively straightforward. However, the final oxidation of amine 32 to (+)-plicamine required significant development. This was eventually achieved using CrO_3 and 3,5-dimethylpyrazole followed by scavenging with Amberlyst 15 resin. The chromium salts were efficiently removed by filtration through a mixed bed containing Varian Chem Elut CE 1005 and Montmorillonite K10 clay to give (+)-plicamine.

The route to plicamine was also be diverted at an earlier stage to afford two further natural products, plicane and obliquine. The ability to divert material in this way is attractive and was facilitated by using immobilized reagent methods. These more advanced syntheses clearly illustrated the opportunities created by embracing these methods for multi-step transformations.

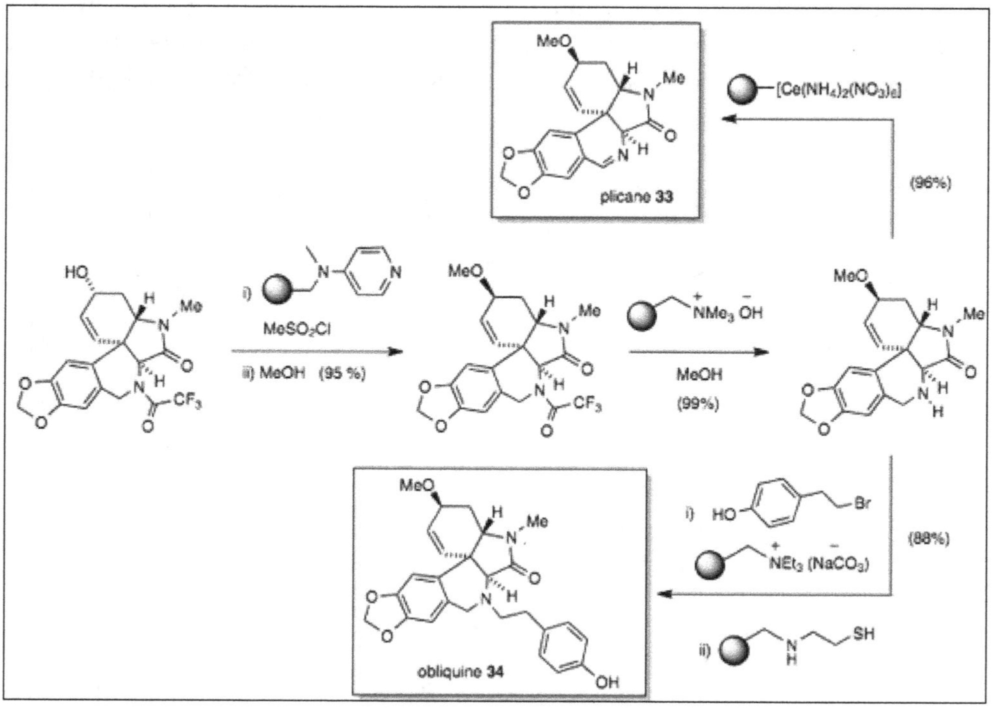

Carpanone

The deceptively complex natural product carpanone was made from commercially available sesamol in a relatively simple set of reactions using immobilized reagents.

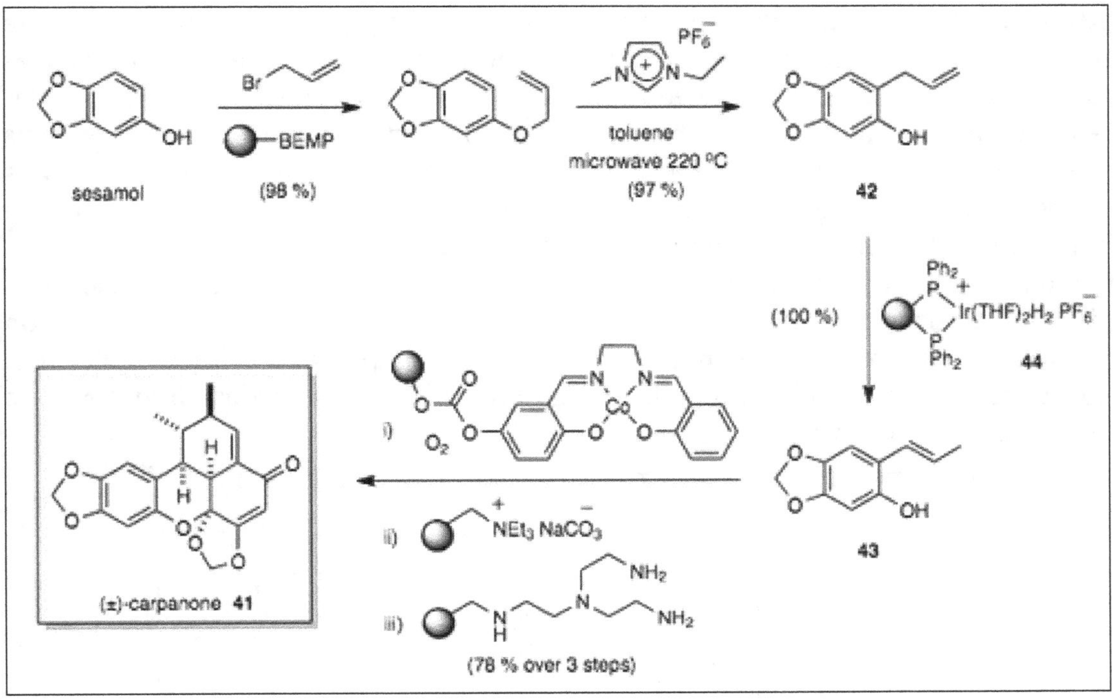

After allylation of sesamol using allyl bromide and PS-BEMP, the product underwent an extremely clean Claisen rearrangement using a combination of toluene and an ionic liquid to absorb the energy from an external microwave source. These binary conditions were simple to operate, since after heating then cooling, the mixture was separated using a simple liquid handler to remove the product-containing toluene layer. The ionic liquid can also be recovered and re-used in further experiments. The toluene fraction containing 42 was then used directly in the double bond isomerisation reaction to yield conjugated compound 43. This was achieved stereoselectively (11:1 trans:cis) by use of a new immobilised iridium catalyst 44. However, it has also been shown to be general for other double bond isomerisations at room temperature. Lastly following original work by Chapman using other oxidants, the phenolic styrene was converted to carpanone by application of a modified Jacobsen Salen cobalt complex under catalytic conditions with molecular oxygen. After scavenging with a carbonate resin to remove unreacted phenols and a trisamine resin to remove an unwanted aldehyde by-product, the crystalline natural product was obtained in excellent yield and purity. Mechanistically, carpanone was finally formed by oxidative dimerisation through carbon coupling of the phenol followed by a highly stereoselective intramolecular Diels-Alder reaction.

Epothilones

The epothilones have generated wide interest in the broad scientific community owing to their ability to inhibit tumour cell proliferation by inducing mitotic arrest through microtubules stabilization. Owing to this activity, they have also become principal targets for many synthesis groups. Indeed they provide an excellent platform to explore the full armoury of supported reagents and scavengers for complex molecule synthesis. The epothilones lend themselves well to convergent synthetic approaches as they can be easily disconnected to more manageable fragments for analogue development; a crucial aspect of pharmaceutical drug programmes. For the synthesis of epothilone C and epothilone A by epoxidation, a convergent synthesis plan was devised that would require the coupling of three major fragments 68, 69 and 70. An important criterion for this route was conducting the work in an efficient and clean manner using only immobilized reagents, scavengers and catch-and-release techniques. The overall aim was to avoid chromatography, crystallisation, distillation and water washes which are common in conventional approaches to these molecules. Consequently, using only immobilized reagents, scavengers and catch-and-release techniques can achieve this goal.

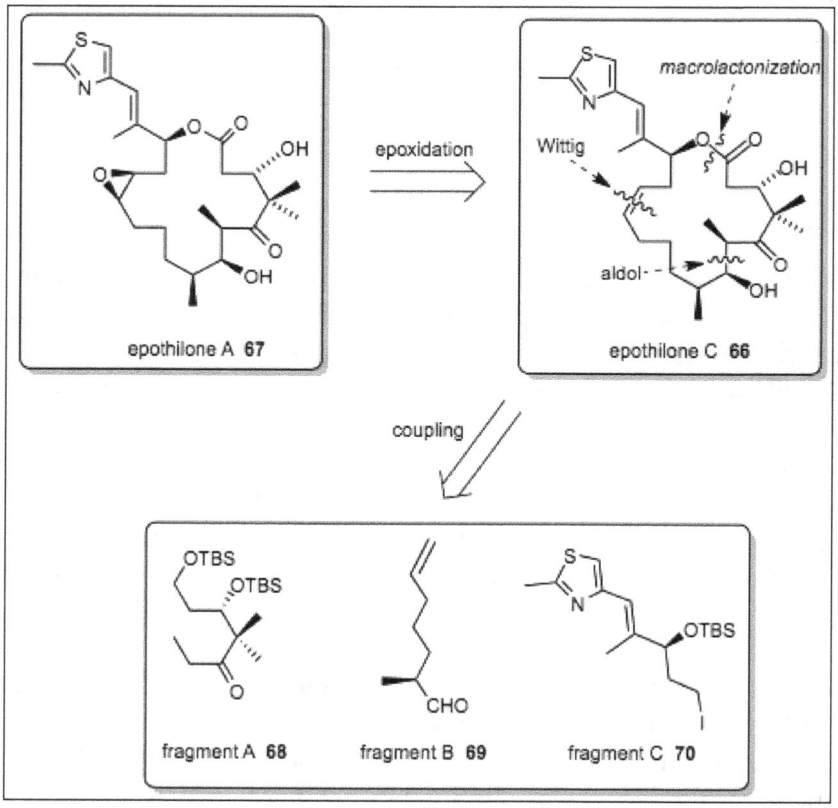

A number of alternative routes to the various fragments were reported. Three routes to fragment A were investigated, but the one shown in below was the shortest, the most efficient, and also proved to be the most easily scaled. The key feature was the formation of the C_2-C_3 bond with concomitant introduction of the desired C_3 stereocentre by application of an asymmetric Mukaiyama aldol reaction. In this synthesis, the aldol reaction progressed in excellent yield to give the alcohol 71 with an enantiomeric excess of 92%. This was achieved using a complex of borane with N-tosylphenylalanine as the chiral ligand. Work-up of this reaction required the addition of a minimum amount of water and a boron selective scavenger Amberlite IRA-743 to quench the reaction and remove the contaminating boric acid. Filtration and solvent removal produced a suspension of amino acid and aldol product 71. However, the insolubility of the N-protected amino acid in non-polar solvents allowed dissolution of the desired aldol product. Subsequent filtration enabled the amino acid to be recovered and recycled whilst concentration of the filtrate gave the purified alcohol 71. After protection as its tert-butyldimethylsilylenol ether 72 and reaction with (trismethylsilylmethyl)lithium followed by a scavenger quench using a carboxylic acid resin, direct filtration and solvent removal yielded the ketone 73. α-Methylation of ketone 73 via its lithium enolate and work-up again with a polymer-supported carboxylic acid gave fragment A in just six steps from commercially available starting material.

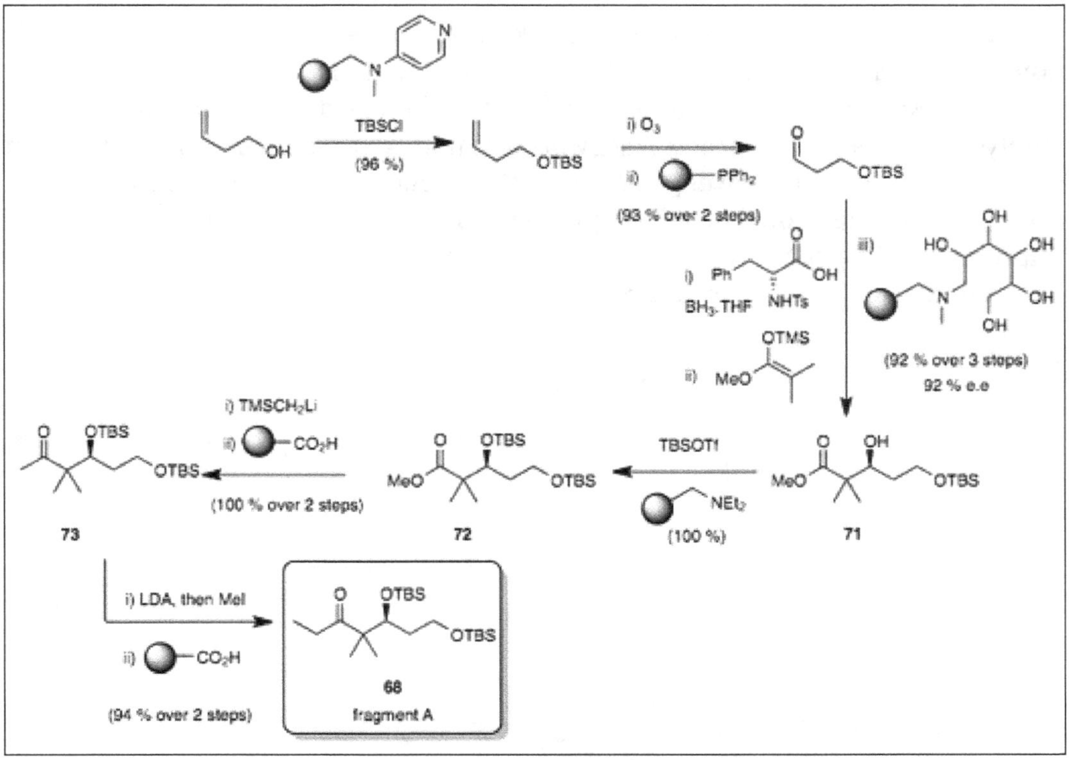

The preparation of the second key coupling partner fragment B was achieved in just five steps from the commercially available bromide (–)-(R)-3-bromo-2-methyl-1-propanol. Protection of the alcohol as its THP-ether using a polymeric sulfonic acid followed by Finkelstein halide exchange gave iodide 75. Homologation of alkyl iodide 75 with a cuprate derived from 3-butenylmagnesium bromide produced the corresponding alkene 76. Addition of a carboxylic acid resin and the trisamine resin quenched the reaction and scavenged dissolved copper salts. Finally after deprotection of the THP acetal, the resulting alcohol was oxidised to the fragment B aldehyde 69. Several oxidants were investigated for this process, but the most expedient proved to be pyridinium chlorochromate PCC on basic alumina.

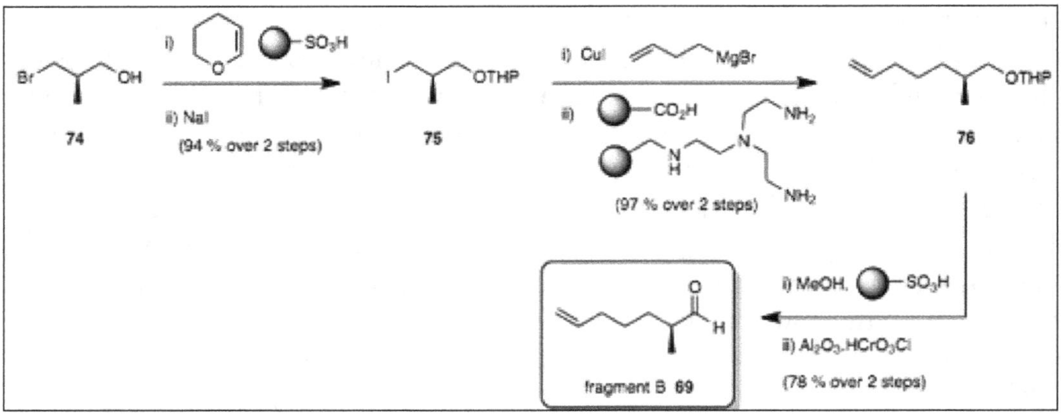

The final fragment C was constructed in a convergent fashion from (S)-α-hydroxyl-g-lactone and the chloromethyl triazole hydrochloride. Lactone 77 was therefore elaborated to ketone 79. Ketone 79 was coupled with a phosphonate derived from thiazole 78 via a highly stereoselective Horner-Wadsworth-Emmons reaction. A polymer-supported aldehyde was used to scavenge any excess phosphonate from this reaction to yield the bis-TBS protected adduct 80. TBS-protected diol 80 was taken through to fragment C. Ultimately iodide 70 was captured onto a polymer-supported triphenylphosphine to produce Wittig salt 81 as the coupling precursor.

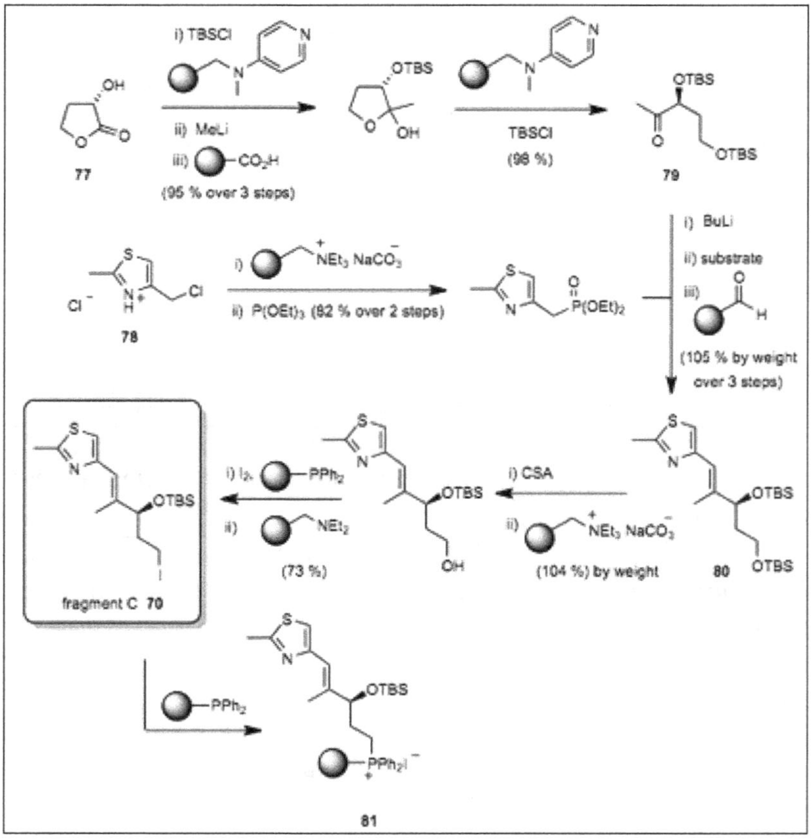

With all the fragments now in hand, the final fusion of the components began. Previously, the stereoselective aldol coupling of fragments A and B to form the C_6-C_7 bond was shown to be highly sensitive to proximal and remote functionality in both fragments. The coupling of ketone 68 with the aldehyde 69 was a novel combination. Using LDA as a base this aldol coupling proceeded in quantitative yield with better than 13:1 stereoselectivity for the desired product. An acetic acid quench was followed by treatment with a diamine functionalised polymer. This resin served two purposes; to remove the excess acid and to sequester a small amount of unreacted aldehyde.

The elaborated adduct 82 was then silyl-protected and the double bond cleaved by ozonolysis. Work-up of the intermediate ozonide was achieved by application of immobilized triphenylphosphine. This is an excellent procedure as the triphenylphosphine oxide produced in normal solution work-ups and in Wittig chemistry causes practical difficulties that often require multiple chromatographic separations to obtain pure material. Here simple filtration was sufficient. The resin bound phosphonium salt 81 derived from fragment C was treated with an excess of sodium hexamethyldisilazide (NaHMDS) followed by washing with anhydrous THF to give an isolable salt-free ylide. This was coupled stereoselectively to give the cis-olefin which required application of a dilute methanolic solution of camphorsulfonic acid to effect selective removal of the primary TBS-protecting group to give the free alcohol 84. This process required a separate scavenging step using an immobilized carbonate resin to remove the acid with the volatile MeOTBS by-product being removed under reduced pressure.

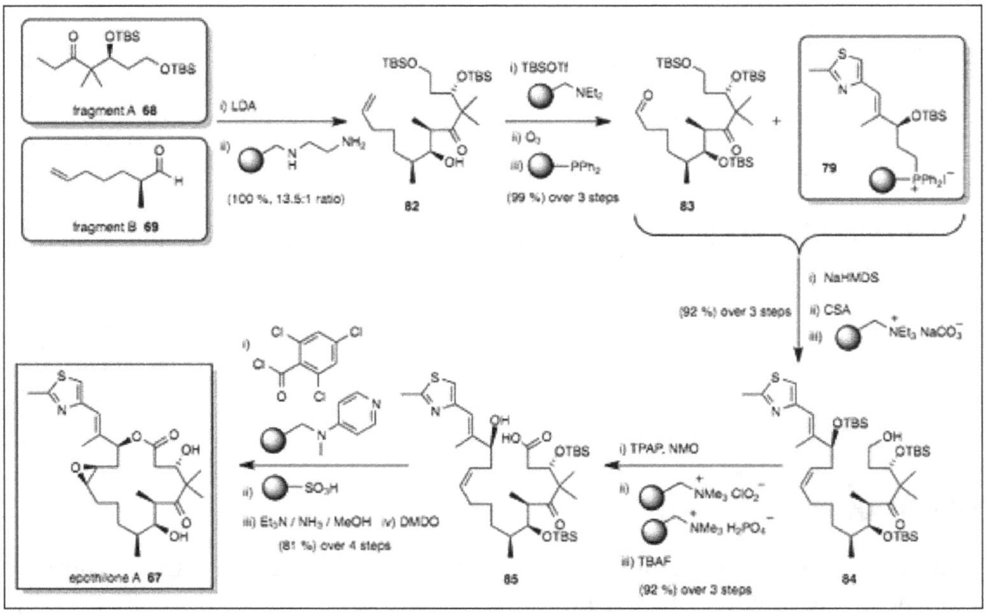

Catalytic tetra-N-propylammonium perruthenate (TPAP) oxidation of alcohol 84 followed by filtration through a pad of silica gel to remove morpholine and ruthenium by-products gave an intermediate aldehyde which was immediately oxidised to the corresponding acid using a previously developed modified Pinnick procedure. Finally, selective desilylation prior to macrocyclisation was finally achieved found using tetrabutyl ammonium fluoride (TBAF) solution to afford intermediate 85. Immobilized versions of fluoride or other acidic resins led to complex mixtures. As a consequence of using the TBAF procedure, an aqueous extraction was necessary, and this gave pure deprotected alcohol 85. This was the first and only water wash used in the whole synthesis to this point. The final steps to epothilone A simply required application of the Yamaguchi macrolactonisation procedure using PS-DMAP and a catch-and-release purification to produce epothilone C. Upon epoxidation with DMDO, epothilone A was obtained.

this synthesis constitutes a triumph for the utility of supported reagents and scavenging techniques for multi-step complex molecule assembly.

Advantages and Disadvantages of Polymer Supported Reagents

The use of polymer supported reagents can combine the benefits of solid – phase chemistry with the advantages of solution – phase synthesis

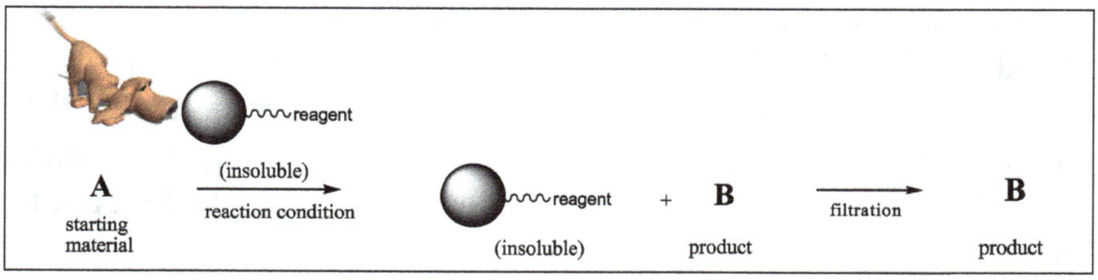

Advantages

- Simplification of reaction work up and product isolation > simple filtrations.
- PSRs can be used in excess without a drawback in the purification step.
- Reactions can be driven to completion more easily than in conventional solution – phase chemistry.
- Potential of recyclization, of higher stability, of reduced toxicity, of simple reaction monitoring by TLC, and simplified automation.
- Multistep procedures in one pot.

Disadvantages

- Supporting the active moiety might reduce its activity due to unfavorable electronic and steric interactions with the support > long reaction times and poor yield.
- The loading of PSRs is in general low.

Green Solvents

Green solvents are environmentally friendly solvents or biosolvents, which are derived from the processing of agricultural crops.

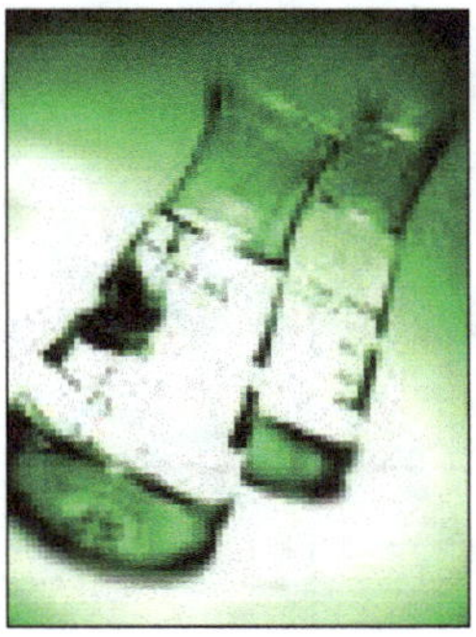

The use of petrochemical solvents are the key to the majority of chemical processes but not without severe implications on the environment. The Montreal Protocol identified the need to re-evaluate chemical processes with regard to their use of volatile organic compounds of VOCs and the impact theses VOCs has on the environment. Green solvents were developed as a more environmentally friendly alternative to petrochemical solvents.

Ethyl lactate, is a green solvent derived from processing corn.

Ethyl lactate is the ester of lactic acid. Lactate esters solvents are commonly used solvents in the paints and coatings industry and have numerous attractive advantages including being 100% biodegradable, easy to recycle, non-corrosive, non-carcinogenic and non-ozone depleting.

Ethyl lactate is a particularly attractive solvent for the coatings industry as a result of its high solvency power, high boiling point, low vapour pressure and low surface tension. It is a desirable coating for wood, polystyrene and metals and also acts as a very effective paint stripper and graffiti remover. Ethyl lactate has replaced solvents including NMP, toluene, acetone and xylene, which has resulted in the workplace being made a great deal safer.

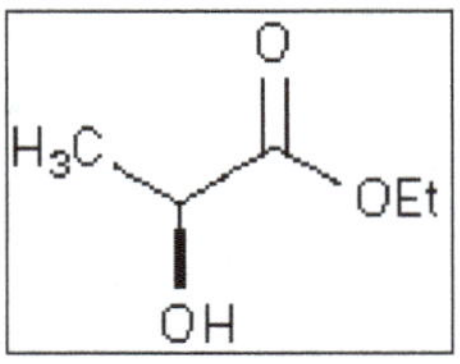

Other applications of ethyl lactate include being an excellent cleaner for the polyurethane industry. Ethyl lactate has a high solvency power which means it has the ability to dissolve a wide range of polyurethane resins. The excellent cleaning power of ethyl lactate also means it can be used to clean a variety of metal surfaces, efficiently removing greases, oils, adhesives and solid fuels. The use of ethyl lactate is highly variable as it has eliminated the use of chlorinated solvents.

The chart shows the synthetic route to products made from ethyl lactate:

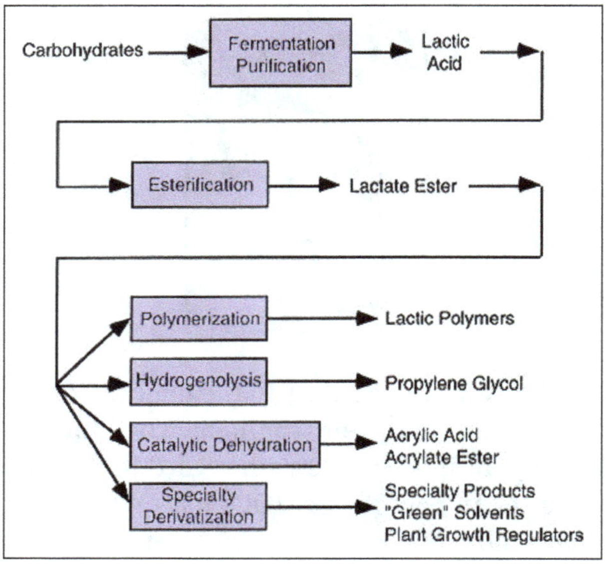

Alternative promising areas of research in the replacement of the use of VOCs in industry include the application of supercritical carbon dioxide and ionic liquids as alternative solvents. Both of which have proved to have valuable applications.

References

- Dimethyl-carbonate, chemistry: sciencedirect.com, Retrieved 8 April, 2020
- Polymer-supported-reagents, synthesis-methodology: leygroup.ch.cam.ac.uk, Retrieved 17 May, 2020
- Kreutzberger., Charles B. (2001), “Chloroformates and Carbonates”, Kirk-Othmer Encyclopedia of Chemical Technology, New York: John Wiley, doi:10.1002/0471238961.0301180204011312.a01.pub2, ISBN 9780471238966
- Bio-solvents-synthesis-industrial-production-and-applications, solvents-ionic-liquids-and-solvent-effects: intechopen.com, Retrieved 17 January, 2020

Green Chemistry and the Environment

6

- Role of Green Chemistry in Controlling Environmental and Ocean Pollution
- Environmental Chemistry
- Environmental Toxicology
- Health Effects of Toxic Chemicals
- Greenhouse Effect
- Environmental Sustainability

Green chemistry aids in reducing pollution by minimizing the hazards of chemical feedstocks, solvents and products. Remediation removes hazardous materials from the environment and maintains environmental sustainability. This chapter closely examines the role of green chemistry in the environment for an extensive understanding of the subject.

Role of Green Chemistry in Controlling Environmental and Ocean Pollution

Green Chemistry controls environmental pollution by using a variety of green alternatives to conventional methods.

Ionic Liquids

Ionic liquids are mixtures of anions and cations, molten salts with melting point around

100 degree-Celsius, which can be used as alternative solvents in organic synthesis. They are very promising as alternatives even to organic solvents as their properties can be tuned by proper selection of cation and anion constituents. For example:

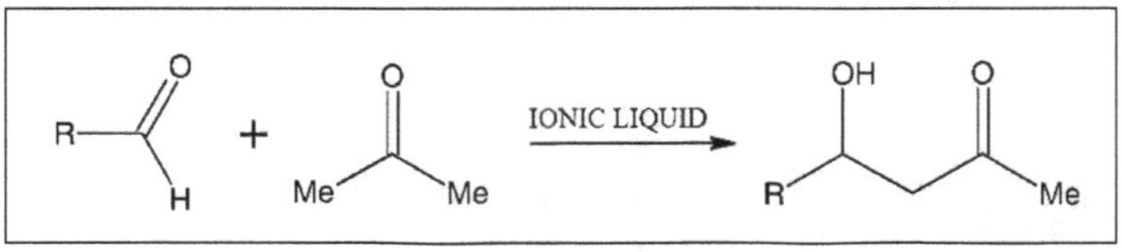

Simple Aldol reaction using Ionic liquids as solvents.

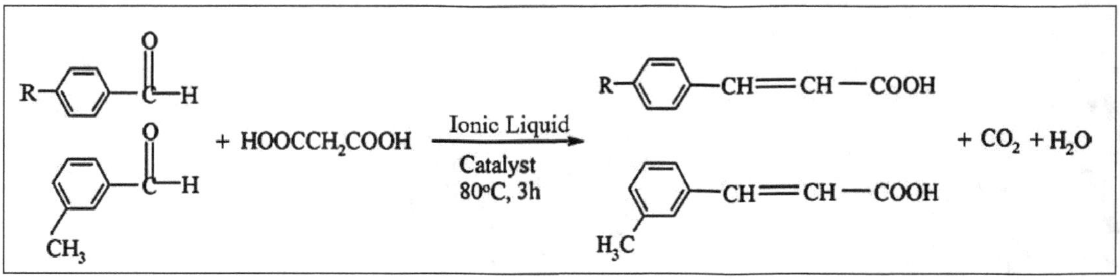

IL catalysed Doebner condensation.

In Doebner condensation, ionic liquids $[Bmim]BF_4$ and $[Bpyr]BF_4$ are used to synthesize $\alpha\beta-$unsaturated carboxylic acid.

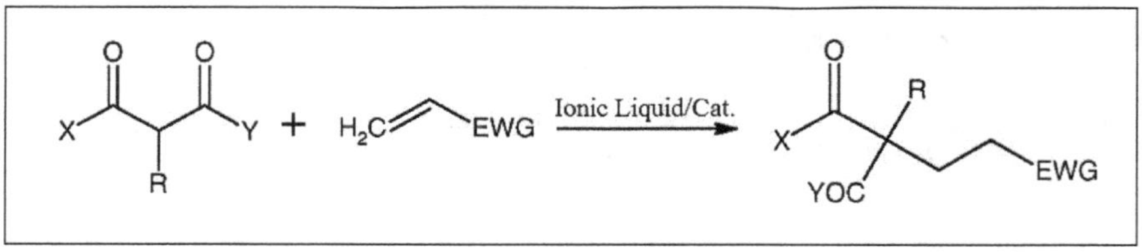

Michael reaction employing a variety of catalysts.

Organic Synthesis in Water

Although water is considered a problem for organic synthesis and the purification process and drying in final products is very cumbersome, yet in recent years, water is considered as good solvent for organic reactions. For example- Synthetic routes of the Diels-Alder reactions in which the hydrophobic properties of some reagents make water an ideal solvent. Water as a solvent accelerates some reactions because some reagents are not soluble and provide selectivity.

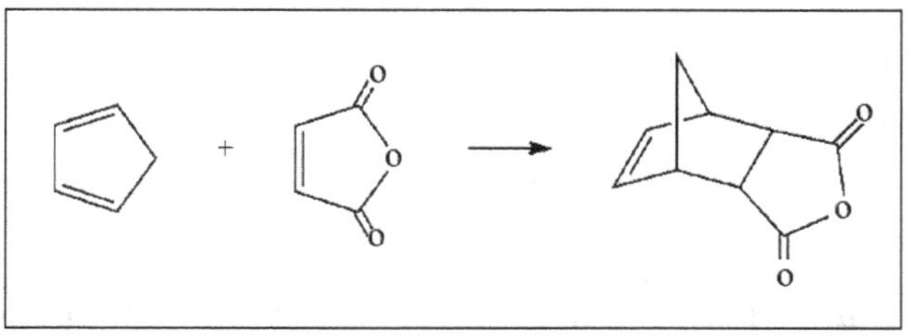

Diels-Alder Reaction.

Super-critical Carbon Dioxide and Super-critical Water

A Super-critical liquid/gas is a liquid/gas at a temperature and pressure above its critical point, where distinct liquid and gas phases do not exist. The supercritical liquid can effuse through solids like a gas and dissolve materials like a liquid. In addition, close to the critical point, small changes in pressure or temperature result in large changes in density allowing many properties of a super-critical liquid to be fine tuned. Supercritical liquids are suitable as a substitute for organic solvents in a range of industrial and laboratory processes. Carbon-Dioxide and Water are the most commonly used supercritical fluids. They are known as "Green Solvents" in many industrial processes, providing high yields in many reactions. Supercritical CO_2 has been tested in various industrially important reactions, such as alkylations, hydroformylations, and hydrogenation, as an alternative reaction medium. Applications of supercritical CO_2 in polymers has been found in polymerization, polymer composite production, polymer blending, particle production, and microcellular foaming. The widest application of supercritical CO_2 has been observed in extraction.

Solvent Free Reactions

In most organic reactions liquid solvents are used such as hydrocarbons, chlorinated hydrocarbons, esters, alcohols, ethers, ammonia, carbon disulphide, water etc. based upon its physical and chemical properties, its suitability to the chemical reaction. But the attempt to develop environment friendly, synthetic procedures has projected the need to minimize use of solvents (VOC) which are the major cause of pollution. Some of solvent free reactions such as Tishchenko reaction, Reformatsky and Luche reaction, Oxidative Coupling reaction of phenols with $FeCl_3$, Solvent free synthesis of Chalcones, etc. are given below.

$$2RCHO \xrightarrow{\text{Catalyst}} RCOOCH_2R$$

Tischchenko reaction.

$$RCHO + BrCH_2COOEt \underset{NH_4Cl}{\overset{Zn}{\rightarrow}} ArCH(OH)CH_2COOEt$$

Reformatsky Reaction.

$$RCHO + BrCH_2 = CH_2 \underset{NH_4Cl}{\overset{Zn}{\rightarrow}} RCH(OH)CH_2CH = CH_2$$

Luche Reaction.

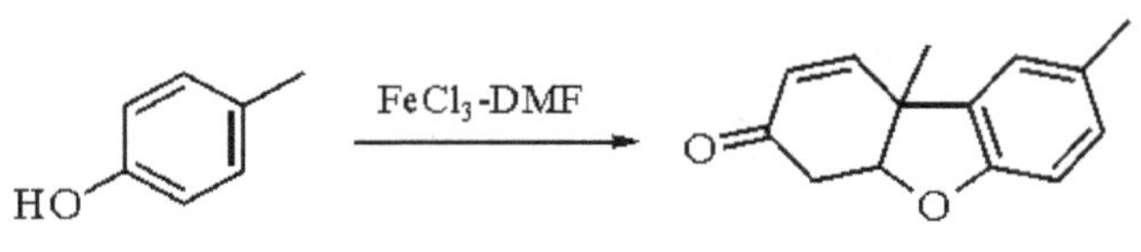

(2)

Oxidative Coupling Reaction of phenols with FeCl.

$$C_6H_5CHO + C_6H_5COCH_3 \xrightarrow{NaOH} C_6H_5-CH=CH-CO-C_6H_5$$

Synthesis of Chalcones.

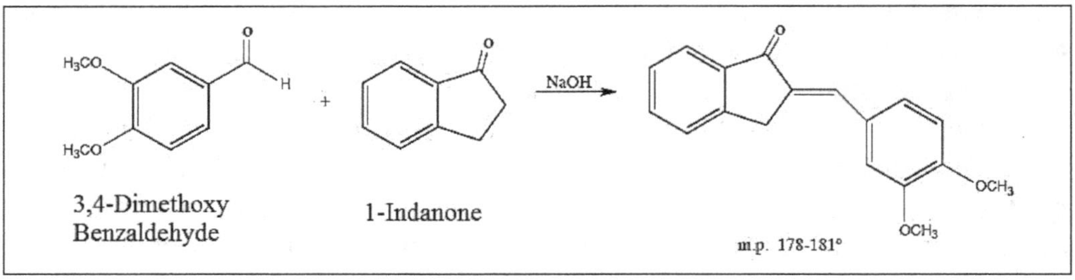

Solvent free aldol condensation.

$$C_6H_5CHO \xrightarrow{Thiamine} C_6H_5COCH(OH)C_6H_5$$

Solvent free Benzoin condensation.

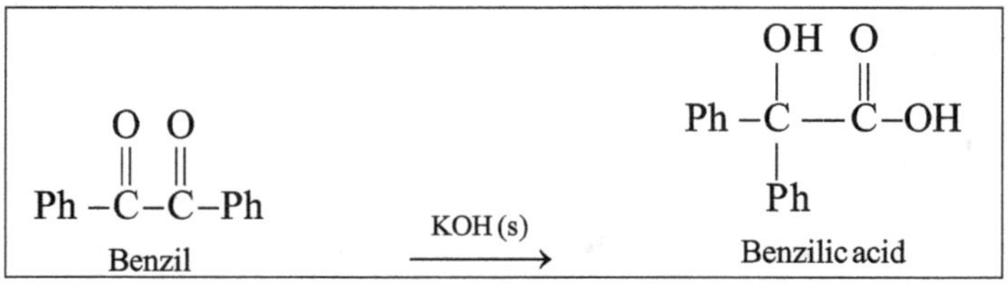

Solvent free conversion of Benzil to Benzilic acid.

The conversion takes place in half an hour on heating the contents on water bath. The same reaction occurs only in one minute by heating in microwave at 540 watt.

Microwave Assisted Solvent Free Organic Synthesis

This technique does not require solvents and are considered "greener" than the conventional methods. Microwave-irradiation in the solid-state is a technique that is being utilized to affect chemical transformations rapidly in contrast to those that have been classically conducted in liquid solutions. The microwaves assisted solvent free reactions can be done in open vessels which prevent the risk of high pressure development within the container and cause explosion. Solvent free reactions also reduce waste discharge. For example microwave-assisted solvent-free preparation of quinoline derivatives by Friedlander coupling condensation between an acetophenone and a 2- aminoacetophenone in the presence of diphenylphosphate (0.1–0.5 equiv.) within 4 min. This procedure afforded product yields of up to 85%, whereas the yield obtained with classical heating under similar conditions did not exceed 24%.

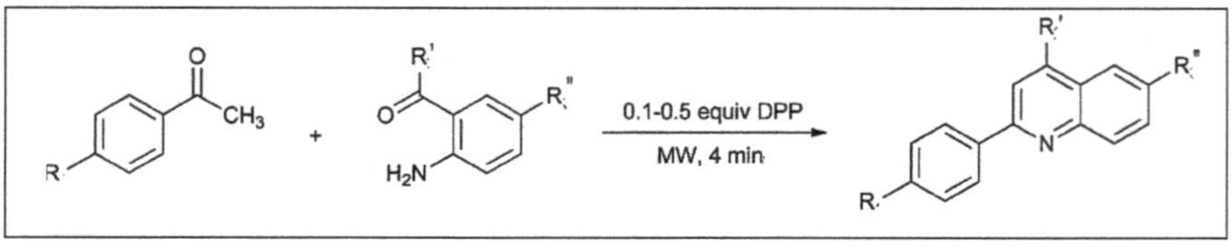

Preparation of quinoline derivatives under microwave in the absence of solvent.

$$C_6H_5COCH(OH)C_6H_5 \xrightarrow[MW]{Zeolite\ A} C_6H_5COCOC_6H_5$$

Oxidation of Benzoin to Benzil without any oxidizing agent.

Green Organic Analysis

In conventional organic analysis, organic compound is fused with sodium metal which is very hazardous as may cause fire if come in contact with moisture or water. Sometimes it violently explodes and even strikes the eyes of the students causing damage. A safe and non-hazardous procedure, which can be alternatively performed, has been proposed to use zinc dust and sodium bicarbonate instead of metallic sodium.

Green Qualitative Inorganic Analysis

In classical technique H_2S is used, which is highly toxic and has adverse effects on humans and the environment. Increasing awareness concerning a healthy environment. NH_4^+ and K^+ are detected in group zero by applying direct tests. Pb^{2+} and Ag^+ are detected by using dilute HCl as their chlorides in group I. Ca^{2+}, Sr^{2+}, Ba^{2+}, and Pb^{2+} are precipitated as sulphates using Na_2SO_4(aq) and ethanol in group II. Cu^{2+}, Cd^{2+}, Fe^{3+}, Mn^{2+}, Co2+, Ni^{2+}, and Mg^{2+} are precipitated in group III as hydroxides using NaOH and H_2O_2. The precipitate is treated with NH_3(aq) which dissolves Cu^{2+}, Cd^{2+}, Ni^{2+}, Co^{2+} forming soluble ammines and are grouped as group IIIB leaving behind a residue containing Fe^{3+}, Mn^{2+}, Mg^{2+} as hydroxides and are grouped as group IIIA. Al^{3+}, Zn^{2+}, Sn^{2+}/Sn^{4+} as soluble hydroxo complexes and Cr^{3+} as CrO_4^{2-} are detected in group IV. Also, the amount of reagents used and thus environmental pollution can be reduced by performing spot tests for cations and anions instead of conventional tests.

Green Quantitative Analysis using Flower Petal Extracts as Indicators

The synthetic indicators like phenolphthalein, methyl orange and phenol red are not only hazardous to health but are also prominent pollutants. The Green Chemistry has proved that these unsafe chemicals can be substituted by the petal extract as an indicator for acid base titration. The accuracy of the observed results has been examined by performing titration between different acids and bases of varying normality using petal extracts which are neither harmful to the environment nor it causes any health hazards e.g. Delonix regia flower petals, Urena lobata, Hibiscus rosa sinensis, Dahlia pinnata etc.

Chemists are also trying to find more environmental friendly and sustainable methods to be used in daily life.

- Biodegradable Plastics: Non-biodegradable plastics are making a giant heap of waste material on earth. Minnesota makes food containers from polylactic acid called as ingeo. Scientist at nature works converted corn starch into a resin that is just as strong as the rigid petroleum based plastic and is currently used for making containers such as water bottles, yogurt pots etc. BASF developed a compostable polyester film called Eco flux they are using for making fully biodegradable bags made from this film with cassava starch and calcium carbonate. Theses bags completely decompose into water, CO_2 and biomass in industrial composting systems. Use of these bags in place of conventional plastics bags, the kitchen and yard waste will quickly vitiate in municipal corporation system.

- Ecofriendly Paint: Oil based 'alkyd 'paints give off large amount of volatile organic compounds (VOCs) as it dries and cures. These VOC's have many environmental effects. Procter & Gamble and Cork composites & polymers established a mixture of soya oil and sugar to be used in place of petroleum petrochemicals derived paints resins and solvents which reduced the hazardous volatiles by 50%. Chermpol MPS, paint formulation use these bio based sepose oils to replace petroleum based solvents and create paint which is safer to use. Sherwin William established water based acrylic alkyd paints from recycled soda bottle plastic (PET), acrylics and soya bean oil. These paints give performance benefits of alkyds and low VOC content of acrylics. In alone 2010 enough quantity of these paints were manufactured to eliminate 362,874Kg of VOC's.

- Green Bleaching agents: Conventionally during manufacturing of good quality white paper, lignin from wood used for it, is removed by placing small pieces of wood into a bath of sodium hydroxide and sodium sulphide followed by its reaction with chlorine. Chlorine during the process also reacts with aromatic rings of the lignin to form chlorinated dioxins and chlorinated furans. These compounds being carcinogens, cause health problems. Terrence Collins of cambegie mellon university developed a green bleaching agent which involves use of H_2O_2 as a bleaching agent in the presence of some activators such as TAML which catalysis the fast conversion of H_2O_2 into hydroxyl radicals that cause bleaching. This bleaching agent breaks down lignin in a shorter time and at much lower temperature.It can be used in laundry and results in lesser use of water.

- Putting out fires the green way: The conventionally used chemical firefighting foams used worldwide discharge toxic substances into environment contaminating water and deleting ozone layer. A new Foam called pyro cool has now been invented to put out fires effectively without producing toxic substances as in other firefighting materials.

- Green dry cleaning of clothes: Percholoroethylene (perc) is the solvent most commonly used in dry cleaning clothes. Perc ($Cl_2C = CCl_2$) is suspected to be carcinogenic and it contaminates ground water on its disposal. A new technology known as micell technology is developed by Joseph De Simons, Timothy Remark and James Mc clain in which liquid carbon dioxide can be used as a safer solvent along with a surfactant to dry clean clothes. This method is now being used commercially by some dry cleaners. Dry cleaning machines have been modified for using this technology so carcinogen PERC is replaced by green solvent.

- Turning turbid water clear in green way: Conventionally, municipality and industrial waste water is made clear by the use of Alum. Alum is found to increase toxic ions in treated water which causes Alzheimer's disease. The tamarind seed kernel powder has been found as effective and economic agent to make waste water clear as alum.
- Biofuels: It can be obtained from biomass which may be obtained from sugar cane, corn, rapeseed, straw, wood, animal and agriculture residues. For example: Bio-diesel which is produced from oil or fat by a process called " transesterification" when burnt in diesel engine with hydrocarbons is found to reduce the petroleum fuel consumption as well as the generation of toxic gaseous substances.
- .Lighter vehicles preparation, reduced costs and carbon dioxide emission: The modern synthetic polyesters has been found to reduce the quantity of foam used in the seats of the car, decreasing its weight sufficiently which decreases the fuel consumption and carbon dioxide emission into atmosphere.
- New Lighting technologies: Organic light emitting diodes (OLEDS) an example of new lighting technologies produce more light with lower energy consumption.
- Use of Unleaded petrol: The higher the octane number better is the quality of petrol. Octane number of petrol these days is increased without addition of lead components such as tetra ethyl lead (TEL).Unleaded petrol is obtained by adding methyl tertiary butyl ether (MTBE) which supplies oxygen of petrol and hence reducing the formation of per-oxy compounds.

Use of Nonpetroleum Fuels

- Power Alcohol: When ethyl alcohol is used in an internal combustion engines, it is called " power alcohol ". It is mixed with Gasoline in the ratio of 4:1 in order to increase its octane number. As in India ethyl alcohol is prepared from molasses, a dark brown mother liquor residue left after the crystallization of sugar thus an enormous amount of residue is consumed and hence reduces the pollution.
- Benzol: It is obtained as side product during coal carbonization. It is also obtained from the fractional distillation of light oil. It is a mixture of benzene (70%), Toluene (18%) and Xylenes (6%) with some other hydrocarbons. It can be used as a component of motor fuels due to its high anti knocking value and hence it reduces the fuels consumption as well as generation of toxic pollutants.

Environmental Chemistry

Environmental chemistry is the scientific study of the chemical and biochemical phenomena that occur in natural places. It should not be confused with green chemistry, which seeks to reduce potential pollution at its source. It can be defined as the study of the sources, reactions, transport, effects, and fates of chemical species in the air, soil, and water environments; and the effect of human activity and biological activity on these. Environmental chemistry is an interdisciplinary science that includes atmospheric, aquatic and soil chemistry, as well as heavily relying on analytical chemistry and being related to environmental and other areas of science.

White bags filled with contaminated stones line the shore near an industrial oil spill.

Environmental chemistry involves first understanding how the uncontaminated environment works, which chemicals in what concentrations are present naturally, and with what effects. Without this it would be impossible to accurately study the effects humans have on the environment through the release of chemicals.

Environmental chemists draw on a range of concepts from chemistry and various environmental sciences to assist in their study of what is happening to a chemical species in the environment. Important general concepts from chemistry include understanding chemical reactions and equations, solutions, units, sampling, and analytical techniques.

Contamination

A contaminant is a substance present in nature at a level higher than fixed levels or that would not otherwise be there. This may be due to human activity and bioactivity. The term contaminant is often used interchangeably with *pollutant*, which is a substance that has a detrimental impact on the surrounding environment. Whilst a contaminant is sometimes defined as a substance present in the environment as a result of human activity, but without harmful effects, it is sometimes the case that toxic or harmful effects from contamination only become apparent at a later date.

The "medium" (e.g. soil) or organism (e.g. fish) affected by the pollutant or contaminant is called a *receptor*, whilst a *sink* is a chemical medium or species that retains and interacts with the pollutant e.g. as carbon sink and its effects by microbes.

Environmental Indicators

Chemical measures of water quality include dissolved oxygen (DO), chemical oxygen demand (COD), biochemical oxygen demand (BOD), total dissolved solids (TDS), pH, nutrients (nitrates and phosphorus), heavy metals, soil chemicals (including copper, zinc, cadmium, lead and mercury), and pesticides.

Applications

Environmental chemistry is used by the Environment Agency in England, Natural Resources Wales, the United States Environmental Protection Agency, the Association of Public Analysts, and other environmental agencies and research bodies around the world to detect and identify the nature and source of pollutants. These can include:

- Heavy metal contamination of land by industry. These can then be transported into water bodies and be taken up by living organisms.
- PAHs (Polycyclic Aromatic Hydrocarbon) in large bodies of water contaminated by oil spills or leaks. Many of the PAHs are carcinogens and are extremely toxic. They are regulated by concentration (ppb) using environmental chemistry and chromatography laboratory testing.
- Nutrients leaching from agricultural land into water courses, which can lead to algal blooms and eutrophication.
- Urban runoff of pollutants washing off impervious surfaces (roads, parking lots, and rooftops) during rain storms. Typical pollutants include gasoline, motor oil and other hydrocarbon compounds, metals, nutrients and sediment (soil).
- Organometallic compounds.

Methods

Quantitative chemical analysis is a key part of environmental chemistry, since it provides the data that frame most environmental studies.

Common analytical techniques used for quantitative determinations in environmental chemistry include classical wet chemistry, such as gravimetric, titrimetric and electrochemical methods. More sophisticated approaches are used in the determination of trace metals and organic compounds. Metals are commonly measured by atomic spectroscopy and mass spectrometry: Atomic Absorption Spectrophotometry (AAS) and Inductively Coupled Plasma Atomic Emission (ICP-AES) or Inductively Coupled Plasma Mass Spectrometric (ICP-MS) techniques. Organic compounds, including PAHs, are commonly measured also using mass spectrometric methods, such as Gas chromatography-mass spectrometry (GC/MS) and Liquid chromatography-mass spectrometry (LC/MS). Tandem Mass spectrometry MS/MS and High Resolution/Accurate Mass spectrometry HR/AM offer sub part per trillion detection. Non-MS methods using GCs and LCs having universal or specific detectors are still staples in the arsenal of available analytical tools.

Other parameters often measured in environmental chemistry are radiochemicals. These are pollutants which emit radioactive materials, such as alpha and beta particles, posing danger to human health and the environment. Particle counters and Scintillation counters are most commonly used for these measurements. Bioassays and immunoassays are utilized for toxicity evaluations of chemical effects on various organisms. Polymerase Chain Reaction PCR is able to identify species of bacteria and other organisms through specific DNA and RNA gene isolation and amplification and is showing promise as a valuable technique for identifying environmental microbial contamination.

Environmental Toxicology

Environmental toxicology is a multidisciplinary field of science concerned with the study of the harmful effects of various chemical, biological and physical agents on living organisms. Ecotoxicology is a subdiscipline of environmental toxicology concerned with studying the harmful effects of toxicants at the population and ecosystem levels.

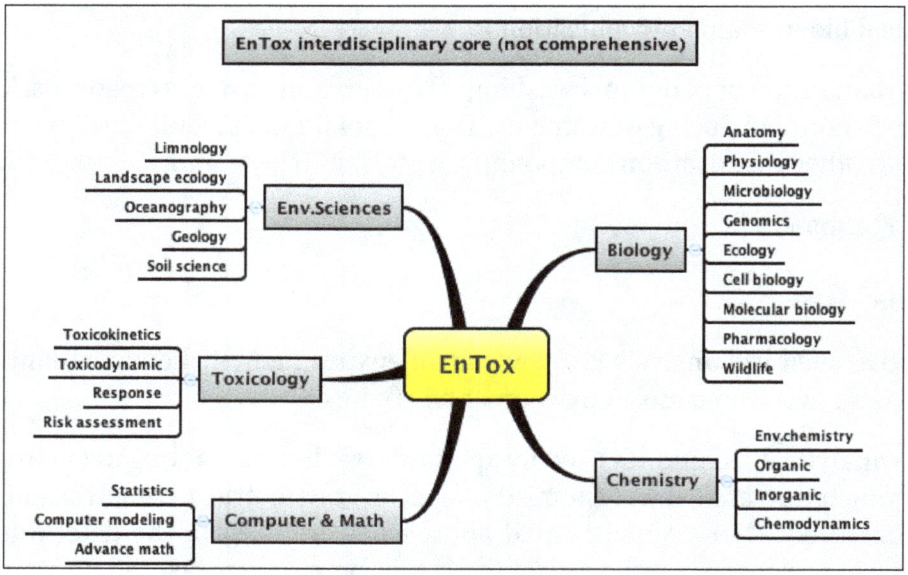

Overview of the interdisciplinarity of environmental toxicology.

Rachel Carson is considered the mother of environmental toxicology, as she made it a distinct field within toxicology in 1962 with the publication of her book Silent Spring, which covered the effects of uncontrolled pesticide use. Carson's book was based extensively on a series of reports by Lucille Farrier Stickel on the ecological effects of the pesticide DDT.

Organisms can be exposed to various kinds of toxicants at any life cycle stage, some of which are more sensitive than others. Toxicity can also vary with the organism's placement within its food web. Bioaccumulation occurs when an organism stores toxicants in fatty tissues, which may eventually establish a trophic cascade and the biomagnification of specific toxicants. Biodegradation releases carbon dioxide and water as by-products into the environment. This process is typically limited in areas affected by environmental toxicants.

Harmful effects of such chemical and biological agents as toxicants from pollutants, insecticides, pesticides, and fertilizers can affect an organism and its community by reducing its species diversity and abundance. Such changes in population dynamics affect the ecosystem by reducing its productivity and stability.

Although legislation implemented since the early 1970s had intended to minimize harmful effects of environmental toxicants upon all species.

Sources of Environmental Toxicity

There are many sources of environmental toxicity that can lead to the presence of toxicants in our food, water and air. These sources include organic and inorganic pollutants, pesticides and biological agents, all of which can have harmful effects on living organisms. There can be so called point sources of pollution, for instance the drains from a specific factory but also non-point sources (diffuse sources) like the rubber from car tires that contain numerous chemicals and heavy metals that are spread in the environment.

PCBs

Polychlorinated biphenyls (PCBs) are organic pollutants that are still present in our environment today, despite being banned in many countries, including the United States and Canada. Due to the persistent nature of PCBs in aquatic ecosystems, many aquatic species contain high levels of this chemical. For example, wild salmon (Salmo salar) in the Baltic Sea have been shown to have significantly higher PCB levels than farmed salmon as the wild fish live in a heavily contaminated environment.

PCBs pertains to a group of human-produced "organic chemicals known as Chlorinated hydrocarbons" The chemical and physical properties of a PCS determine the quantity and location chlorine and unlike other chemicals, they have no form of identification.The range of toxicity is not consistent and because PCBs have certain properties (chemical stability, non-flammability) they have been used in a colossal amount of commercial and industrial practices. Some of those include, "Electrical, heat transfer and hydraulic equipment, plasticizers in paints, plastics and rubber products and pigments, dyes and carbonless copy paper" to name a few.

Heavy Metals

Heavy metals found in food sources, such as fish can also have harmful effects. These metals can include mercury, lead and cadmium. It has been shown that fish (i.e. rainbow trout) are exposed to higher cadmium levels and grow at a slower rate than fish exposed to lower levels or none. Moreover, cadmium can potentially alter the productivity and mating behaviours of these fish. Heavy metals can not only affect behaviors, but also the genetic makeup in aquatic organisms. Researchers wanted to determine as to what effect metal contamination had on evolutionary responses among populations of yellow perch. Along the gradient, genetic diversity over all loci was negatively correlated with liver cadmium contamination. Additionally, there was a negative correlation observed between copper contamination and genetic diversity. Some aquatic species have evolved heavy metal tolerances. In response to high heavy metal concentrations a Dipteran species, Chironomus riparius, of the midge family, Chironomidae, has evolved to become tolerant to Cadmium toxicity in aquatic environments. Altered life histories, increased Cd excretion, and sustained growth under Cd exposure is evidence that shows that Chironomus riparius exhibits genetically based heavy metal tolerance.

Metals Toxicity

The most known or common types of heavy metals include zinc, arsenic, copper, lead, nickel, chromium and cadmium. All of these types cause certain risks on human and environment health.

Though certain amount of these metals can actually have an important role in, for example, maintaining certain biochemical and physiological, "functions in living organisms when in very low concentrations, however they become noxious when they exceed certain threshold concentrations." Heavy metal are a huge part of environmental pollutions and their toxicity "is a problem of increasing significance for ecological, evolutionary, nutritional and environmental reasons."

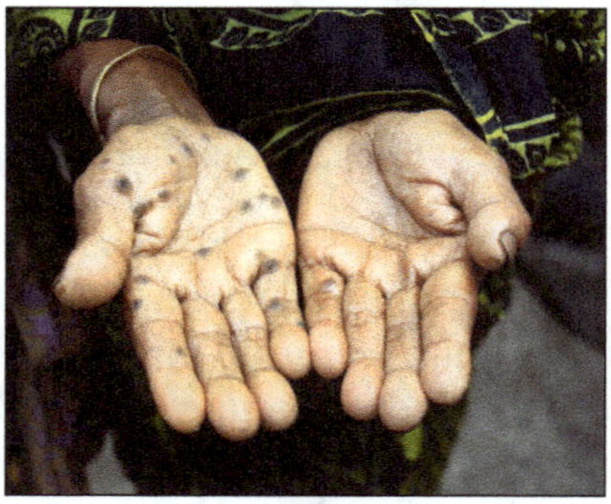

Person has been exposed to Arsenic Poisoning through contaminated water.

Arsenic

Arsenic, one of the most important heavy metals, causes health problems within ecological and humans. It is, "semimetallic property, is prominently toxic and carcinogenic, and is extensively available in the form of oxides or sulfides or as a salt of iron, sodium, calcium, copper, etc." Not only that, but it is also the most abundant element here on earth and its specific inorganic forms are very dangerous to living creatures (animals, plants, and humans) and the environment.

Effects of Arsenic on humans is that they can cause cancer in the bladder, skin, lungs and liver. One of the major areas in which humans are exposed to arsenic is through contaminated water which is a problem in more than 30 countries in the world.

Humans tend to encounter arsenic by, "natural means, industrial source, or from unintended sources." Water can become contaminated by arsenical pesticides or just natural arsenical chemicals. There are some cases in which arsenic has been used in suicidal attempts and can result in acute poisoning. Arsenic, "is a protoplastic poison since it affects primarily the sulphydryl group of cells causing malfunctioning of cell respiration, cell enzymes and mitosis."

Lead

Another extremely toxic metal, lead can and has been known to cause, "extensive environmental contamination and health problems in many parts of the world." The physical appearance of lead is bright and silver colored metal. Some sources of lead pollution in the environment include Metal plating and fishing operations, soil waste, factory chimneys, smelting of ores, wastes from batter industries, fertilizers and pesticides and many more. Unlike, other metals such as copper, lead only plays a physiological aspect and no biological functions. "more than 100 to 200,000 tons of lead per year is being released from vehicle exhausts" and some can be brought in by plants, flow in water or fixation into the soil.

Humans come in contact with lead through mining, fossil fuel burning. In burning, lead and its compounds are exposed into air, soil, and water. Lead can have different effects on the body and effects the central nervous system. Someone who has come in contact with lead can have either acute or chronic lead poisoning. Those who experience acute poisoning have symptoms such as, appetite, headache, hypertension, abdominal pain, renal dysfunction, fatigue, sleeplessness, arthritis, hallucinations and vertigo." Chronic exposure on the other hand, can cause more severe symptoms such as, "mental retardation, birth defects, psychosis, autism, allergies, dyslexia, weight loss, hyperactivity, paralysis, muscular weakness, brain damage, kidney damage and may even cause death."

Mercury

Equally as toxic as the previous heavy metals, Mercury, A shiny silver-white, can transform into a colorless and odorless gas when heated up. Mercury highly affects the marine environment and there have been many studies conducted on the effects on the water environment. The biggest sources of mercury pollution include, "agriculture, municipal wastewater discharges, mining, incineration, and discharges of industrial wastewater" all relatively connected to water.

Mercury exists in three different forms and all three possess different levels of bioavailability and toxicity. The three forms include, organic compounds, metallic elements and inorganic salts. As stated above, they are present in water resources such as oceans, rivers and lakes. They are absorb by microorganism, and go through, "biomagnification causing significant disturbance to aquatic lives."

Mercury hurts marine life but can also be very hurtful towards humans' nervous system. Higher levels of mercury exposure can change many brain functions. It can, "lead to shyness, tremors, memory problems, irritability, and changes in vision or hearing."

Cadmium

Cadmium is the 7th most toxic heavy metal. Cadmium is interesting in that once it is exposed to humans (at work) or animals in their environment, it will accumulate inside the body throughout the life of the human/animal. Though cadmium was used as replacement for tin in WWI and pigment in paint industries back in the day, now day it is seen mostly in rechargeable batteries, tobacco smoke and some alloys production.

The effects of Cadmium on the kidney and bones is huge. It can cause bone mineralization which "is the process of laying down minerals on a matrix of the bone". This can happen through renal dysfunction or bone damage.

Chromium

The 7th most abundant element, Chromium, can occur naturally when one burns oil and coal and is release into the environment through sewage and fertilizers. Chromium usage can be seen in, "industries such as metallurgy, electroplating, production of paints and pigments, tanning, wood preservation, chemical production and pulp and paper production." Chromium toxicity affects the "biological processes in various plants such as maize, wheat, barley, cauliflower, citrullus and in vegetables. Chromium toxicity causes chlorosis and necrosis in plants."

Pesticides

Pesticides are a major source of environmental toxicity. These chemically synthesized agents have been known to persist in the environment long after their administration. The poor biodegradability of pesticides can result in bioaccumulation of chemicals in various organisms along with biomagnification within a food web. Pesticides can be categorized according to the pests they target. Insecticides are used to eliminate agricultural pests that attack various fruits and crops. Herbicides target herbal pests such as weeds and other unwanted plants that reduce crop production.

DDT

Dichlorodiphenyltrichloroethane (DDT) is an organochlorine insecticide that has been banned due to its adverse effects on both humans and wildlife. DDT's insecticidal properties were first discovered in 1939. Following this discovery, DDT was widely used by farmers in order to kill agricultural pests such as the potato beetle, coddling moth and corn earworm.

DDT is not easily biodegradable and thus the chemical accumulates in soil and sediment runoff. Water systems become polluted and marine life such as fish and shellfish accumulate DDT in their tissues. Furthermore, this effect is amplified when animals who consume the fish also consume the chemical, demonstrating biomagnification within the food web. The process of biomagnification has detrimental effects on various bird species because DDT and DDE accumulate in their tissues inducing egg-shell thinning. Humans who consume animals or plants that are contaminated with DDT experience adverse health effects. Various studies have shown that DDT has damaging effects on the liver, nervous system and reproductive system of humans.

Sulfuryl Fluoride

Sulfuryl fluoride is an insecticide that is broken down into fluoride and sulfate when released into the environment. Fluoride has been known to negatively affect aquatic wildlife. Elevated levels of fluoride have been proven to impair the feeding efficiency and growth of the common carp (Cyprinus carpio). Exposure to fluoride alters ion balance, total protein and lipid levels within these fish, which changes their body composition and disrupts various biochemical processes.

Cyanobacteria and Cyanotoxins

Cyanobacteria, or blue-green algae, are photosynthetic bacteria. They grow in many types of water. Their rapid growth ("bloom") is related to high water temperature as well as eutrophication (resulting from enrichment with minerals and nutrients often due to runoff from the land that induces excessive growth of these algae). Many genera of cyanobacteria produce several toxins. Cyanotoxins can be dermatotoxic, neurotoxic, and hepatotoxic, though death related to their exposure is rare. Cyanotoxins and their non-toxic components can cause allergic reactions, but this is poorly understood. Despite their known toxicities, developing a specific biomarker of exposure has been difficult because of the complex mechanism of action these toxins possess.

Occurrence of Cyanotoxins in Drinking Water

The occurrence of this toxin in drinking water depends on a couple of factors. One, is the drinking water's level in raw source water and secondly, it depends on the effectiveness of removing these toxins from water when drinking water is actually being produced. Due to being no data on the absence/presence of these toxins in drinking water, it is very hard to actually monitor the amounts that are present in finished water.

Effects of Toxins on Humans

Though data on the effects of these two toxins are limited, from what is available it suggests the toxins attack the liver and kidney. There was an hepatoenteritis-like outbreak due to the consumption of water that contained, "*C. raciborskii*, a cyanobacteria that can produce cylindrospermopsin." Most cases (typically involving children) have required they be taken to a hospital. The results of hospitilation include: Vomiting, kidney damage (due to lose of water, protein and electrolytes) fever, bloody diarrhea, and headaches.

Health Effects of Toxic Chemicals

There are many materials used in the workplace that can be hazardous. However, in order for them to affect your health, they must come into contact with the body or be absorbed into the body. When assessing the potential health effects from working with a particular material, it is necessary to understand the difference between "toxicity" and "hazard". Toxicity is the ability of a substance to produce an unwanted effect when the chemical has reached a sufficient concentration at a certain site in the body.

The more toxic a material is, the smaller the amount of it is necessary to be absorbed before harmful effects are caused. The lower the toxicity, the greater the quantity of it is necessary to be absorbed. Generally, the toxicity of a chemical is determined through experiments on animals (quite often rats), and measured in terms of the amount of material necessary to cause death in 50% of the test animals. These values are called LD50 (lethal dose) or LC50 (lethal concentration), and are usually given in weight of material per kg of body weight or airborne concentration of material per set time period respectively.

Hazard is the probability that this concentration in the body will occur.

Toxicity is an inherent property of the material. A material may be very toxic, but not hazardous, if it is handled properly and is not absorbed into the body. On the other hand, a material may have a very low toxicity, but be very hazardous.

Example:

- An open container of an acid is much more hazardous than a closed container of the same material.
- Two liquids may possess the same degree of toxicity but present different degrees of hazard:
- One material may be non-irritating to the eyes and nose and odourless. The other may be irritating to the eyes or respiratory system and possess a pungent odour.

- The latter material, because of its warning properties presents a lesser degree of hazard.

In order for toxicants to affect the human system either they must cause damage to external tissues, such as the skin or eyes, or they must be able to enter the body by some mechanism.

Routes of Entry

There are three primary routes of entry into the body: ingestion, skin or eye absorption, inhalation and injection. Ingestion- This means taking a material into the body by mouth (swallowing). Ingestion of toxic materials may occur as a result of eating in a contaminated work area.

Absorption- Substances that contact the eye and the skin may be either absorbed into the body or cause local effects. For the majority of organic compounds, the contribution from skin absorption to the total exposure should not be neglected.

Inhalation- This means taking a material into the body by breathing it in. In the lungs, very tiny blood vessels are in constant contact with the air we breathe in. As a result, airborne contaminants can be easily absorbed through this tissue. In the occupational environment, this is generally the most important route of entry. Injection- This means taking in biological or chemical substances into the body when the skin is accidentally punctured with a sharp object (e.g. contaminated needle).

Health Effects – Chronic vs. Acute

Once a toxic substance has contacted the body it may have either acute (immediate) or chronic (long term) effects.

Example: Spilling acid on your hand will cause immediate harm, i.e. a burn to the skin.

Exposure to asbestos or tobacco smoke may result in lung cancer after as much as twenty years (this is a long term effect).

Exposure – Chronic vs. Acute

Exposure can be classified as chronic or acute. In chronic exposures, the dose is delivered at some frequency (daily or weekly usually) over a period of time. In acute exposures, the dose is delivered in a single event and absorption is rapid. Usually, a chronic exposure occurs at low concentration and acute exposure at high concentration.

Some materials may only cause harm if given acutely, not having any effect in the long term. Other materials may not exhibit an effect in the short term, but may cause problems after prolonged exposure.

Physiological Classification of Materials

This classification identifies toxic materials on the basis of biologic action.

- Irritants: Refers to some sort of aggravation of whatever tissue the material comes in contact with. e.g. ammonia, nitrogen dioxide.
- Asphyxiants: Exert their effects through a depletion of oxygen to the tissues e.g. simple asphyxiants – carbon dioxide, nitrogen, methane, hydrogen chemical asphyxiants – carbon monoxide, hydrogen cyanide, hydrogen sulphide.
- Narcotics or Anaesthetics: The main toxic action is the depressant effect upon the Central Nervous System. e.g. – many organics, chloroform, xylene.
- Systemic Poisons: The main toxic action includes the production of internal damage e.g. Hepatotoxic agents – toxic effects produce liver damage. eg. carbon tetrachloride.
- e.g. Nephrotoxic agents: Toxic effects produce kidney damage eg. some halogenated hydrocarbons
- Carcinogens: Agents/compounds that will induce cancer in humans. e.g. benzene, arsenic, inorganic salts of chromium, nickel, beryllium.
- Mutagens: Agents that affect the cells of the exposed people in such a way that it may cause cancer in the exposed individual or an undesirable mutation to occur in some later generation. e.g. radiation, variety of chemical agents that alter the genetic message.
- Teratogens: Agents or compounds that a pregnant woman takes into her body that generate defects in the fetus e.g. Thalidomide, possibly steroids
- Sensitizers: Agents that may cause allergic or allergic-like responses to occur.

After an initial exposure to a substance an individual may become sensitized to that substance. Subsequent exposures to the same substance, often at a much lower concentration than before, produces an allergic response. This response may be a skin rash (dermatitis) or an asthmatic-like attack, depending on the route of exposure. e.g. cutting oils, isocyanates in polyurethane foam operations and paint spraying operations, some laboratory solvents.

Greenhouse Effect

Diagram showing light energy (white arrows) emitted by the sun, warming the earth's surface which then emits the energy as heat (orange arrows), which warms the atmosphere and is then re-emitted as heat by three of the greenhouse gas molecules (water, carbon dioxide, and methane).

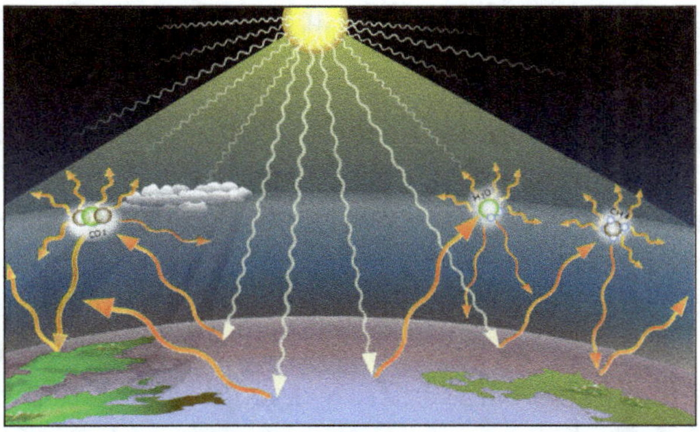

The greenhouse effect is the process by which radiation from a planet's atmosphere warms the planet's surface to a temperature above what it would be without this atmosphere.

Radiatively active gases (i.e., greenhouse gases) in a planet's atmosphere radiate energy in all directions. Part of this radiation is directed towards the surface, warming it. The intensity of the downward radiation – that is, the strength of the greenhouse effect – will depend on the atmosphere's temperature and on the amount of greenhouse gases that the atmosphere contains. Earth's natural greenhouse effect is critical to supporting life, and initially was a precursor to life moving out of the ocean onto land. Human activities, however, mainly the burning of fossil fuels and clearcutting of forests, have accelerated the greenhouse effect and caused global warming.

The planet Venus experienced runaway greenhouse effect, resulting in an atmosphere which is 96% carbon dioxide, with surface atmospheric pressure roughly the same as found 900 m (3,000 ft) underwater on Earth. Venus may have had water oceans, but they would have boiled off as the mean surface temperature rose to 735 K (462 °C; 863 °F). The term "greenhouse effect" continues to see use in scientific circles and the media despite being a slight misnomer, as an atmosphere reduces radiative heat loss while a greenhouse blocks convective heat loss. The result, however, is an increase in temperature in both cases.

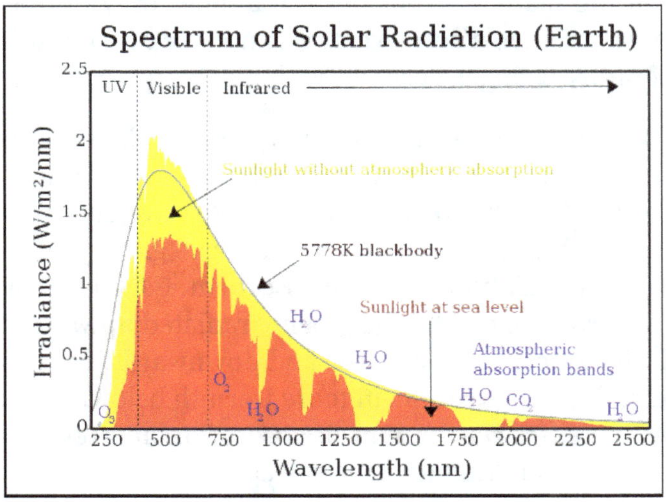

The solar radiation spectrum for direct light at both the top of Earth's atmosphere and at sea level.

Earth receives energy from the Sun in the form of ultraviolet, visible, and near-infrared radiation. About 26% of the incoming solar energy is reflected to space by the atmosphere and clouds, and 19% is absorbed by the atmosphere and clouds. Most of the remaining energy is absorbed at the surface of Earth. Because the Earth's surface is colder than the Sun, it radiates at wavelengths that are much longer than the wavelengths that were absorbed. Most of this thermal radiation is absorbed by the atmosphere and warms it. The atmosphere also gains heat by sensible and latent heat fluxes from the surface. The atmosphere radiates energy both upwards and downwards; the part radiated downwards is absorbed by the surface of Earth. This leads to a higher equilibrium temperature than if the atmosphere did not radiate.

An ideal thermally conductive blackbody at the same distance from the Sun as Earth would have a temperature of about 5.3 °C (41.5 °F). However, because Earth reflects about 30% of the incoming sunlight, this idealized planet's effective temperature (the temperature of a blackbody that would emit the same amount of radiation) would be about −18 °C (0 °F). The surface temperature of this hypothetical planet is 33 °C (59 °F) below Earth's actual surface temperature of approximately 14 °C (57 °F). The greenhouse effect is the contribution of greenhouse gases to this difference.

The idealized greenhouse model is a simplification. In reality the atmosphere near the Earth's surface is largely opaque to thermal radiation and most heat loss from the surface is by convection. However radiative energy losses become increasingly important higher in the atmosphere, largely because of the decreasing concentration of water vapor, an important greenhouse gas. Rather than the surface itself, it is more realistic to think of the greenhouse effect as applying to a layer in the mid-troposphere, which is effectively coupled to the surface by a lapse rate. A simple picture also assumes a steady state, but in the real world, the diurnal cycle as well as the seasonal cycle and weather disturbances complicate matters. Solar heating applies only during daytime. During the night, the atmosphere cools somewhat, but not greatly, because its emissivity is low. Diurnal temperature changes decrease with height in the atmosphere.

Within the region where radiative effects are important, the description given by the idealized greenhouse model becomes realistic. Earth's surface, warmed to an "effective temperature" around −18 °C (0 °F), radiates long-wavelength, infrared heat in the range of 4–100 μm. At these wavelengths, greenhouse gases that were largely transparent to incoming solar radiation are more absorbent. Each layer of atmosphere with greenhouse gases absorbs some of the heat being radiated upwards from lower layers. It reradiates in all directions, both upwards and downwards; in equilibrium the same amount as it has absorbed. This results in more warmth below. Increasing the concentration of the gases increases the amount of absorption and reradiation, and thereby further warms the layers and ultimately the surface below.

Greenhouse gases—including most diatomic gases with two different atoms (such as carbon monoxide, CO) and all gases with three or more atoms—are able to absorb and emit infrared radiation. Though more than 99% of the dry atmosphere is IR transparent (because the main constituents—N_2, O_2, and Ar—are not able to directly absorb or emit infrared radiation), intermolecular collisions cause the energy absorbed and emitted by the greenhouse gases to be shared with the other, non-IR-active, gases.

Greenhouse Gases

By their percentage contribution to the greenhouse effect on Earth the four major gases are:

Atmospheric gases only absorb some wavelengths of energy but are transparent to others. The absorption patterns of water vapor (blue peaks) and carbon dioxide (pink peaks) overlap in some wavelengths. Carbon dioxide is not as strong a greenhouse gas as water vapor, but it absorbs energy in longer wavelengths (12–15 micrometers) that water vapor does not, partially closing the "window" through which heat radiated by the surface would normally escape to space.

- Water vapor, 36–70%
- Carbon dioxide, 9–26%
- Methane, 4–9%
- Ozone, 3–7%

It is not possible to assign a specific percentage to each gas because the absorption and emission bands of the gases overlap (hence the ranges given above). Clouds also absorb and emit infrared radiation and thus affect the radiative properties of the atmosphere.

Role in Climate Change

Strengthening of the greenhouse effect through human activities is known as the enhanced (or anthropogenic) greenhouse effect. This increase in radiative forcing from human activity is attributable mainly to increased atmospheric carbon dioxide levels. According to the 2014 Assessment Report from the Intergovernmental Panel on Climate Change, "atmospheric concentrations of carbon dioxide, methane and nitrous oxide are unprecedented in at least the last 800,000 years. Their effects, together with those of other anthropogenic drivers, have been detected throughout the climate system and are extremely likely to have been the dominant cause of the observed warming since the mid-20th century'". CO_2 is produced by fossil fuel burning and other activities such as cement production and tropical deforestation. Measurements of CO_2 from the Mauna Loa observatory show that concentrations have increased from about 313 parts per million (ppm) in 1960, passing the 400 ppm milestone on May 9, 2013. The current observed amount of CO_2 exceeds the geological record maxima (~300 ppm) from ice core data. The effect of combustion-produced carbon dioxide on the global climate, a special case of the greenhouse effect first described in 1896 by Svante Arrhenius, has also been called the Callendar effect.

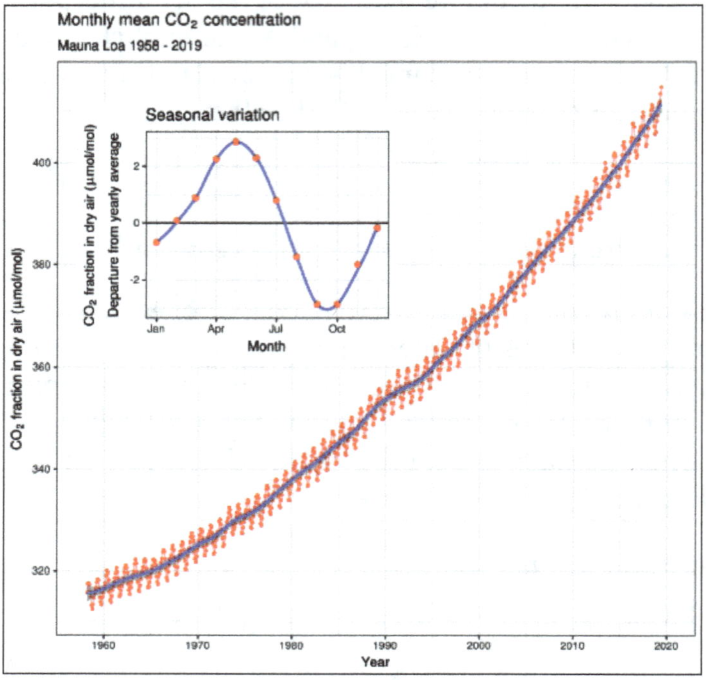

The Keeling Curve of atmospheric CO_2 concentrations.

Over the past 800,000 years, ice core data shows that carbon dioxide has varied from values as low as 180 ppm to the pre-industrial level of 270 ppm. Paleoclimatologists consider variations in carbon dioxide concentration to be a fundamental factor influencing climate variations over this time scale.

Real Greenhouses

A modern Greenhouse.

The "greenhouse effect" of the atmosphere is named by analogy to greenhouses which become warmer in sunlight. However, a greenhouse is not primarily warmed by the "greenhouse effect". "Greenhouse effect" is actually a misnomer since heating in the usual greenhouse is due to the reduction of convection, while the "greenhouse effect" works by preventing absorbed heat from leaving the structure through radiative transfer.

A greenhouse is built of any material that passes sunlight: usually glass or plastic. The sun warms the ground and contents inside just like the outside, and these then warm the air. Outside, the warm air near the surface rises and mixes with cooler air aloft, keeping the temperature lower than inside, where the air continues to heat up because it is confined within the greenhouse. This can be demonstrated by opening a small window near the roof of a greenhouse: The temperature will drop considerably. It was demonstrated experimentally that a (not heated) "greenhouse" with a cover of rock salt (which is transparent to infrared) heats up an enclosure similarly to one with a glass cover. Thus greenhouses work primarily by preventing convective cooling.

Heated greenhouses are yet another matter: As they have an internal source of heating, it is desirable to minimise the amount of heat leaking out by radiative cooling. This can be done through the use of adequate glazing.

Environmental Sustainability

Environmental sustainability is talked about in relation to all aspects of our lives – from creating eco homes and environmentally conscious communities to sourcing sustainable food, renewable energy, low impact furniture and clothing.

Nature has a tremendous capacity to sustain itself. Be it the maintenance of the soil fertility or the process of forestation, when left alone nature can rejuvenate itself to its original self. However, the speed at which the natural resources are getting depleted, it is not possible for nature to sustain itself anymore. To ensure that the ecological balance is not further lost we must make changes in the way we utilize the natural resources for our benefits. We have to employ environmentally sustainable methods to ensure that the future generations can enjoy the natural resources of our planet in the same way as we are utilizing it.

As, a result of which, many steps have been taken towards environmental sustainability to reduce the impact on the environment. Below are a few of the examples of environmental sustainability.

Sustainable Agriculture

Repeated cultivation of the same crops on the same patch of soil can lead to the depletion of the quality of the soil and the quality of the produce. This leads to the use of chemical fertilizers, pesticides and other medicines on plants to improve the productivity. However, this leads to the contamination of the crops with chemicals which can harm human health.

Sustainable agriculture integrates plant and animal production practices that are specific to the site and not only enhances the productivity but also has minimal impact on the environment. It incorporates biological and ecological processes into agriculture reducing the dependency on non-renewable resources and inorganic unsustainable inputs.

Crop rotation is a very effective way to maximize the productivity of the crops. Rotation of crops helps in rejuvenating the fertility of the soil and help to lower the presence of disease in the soil. Animal and crop waste is recycled and used as manure to improve the content of nitrogen in the soil. Water management is another very important factor that contributes to sustainable agriculture. In many droughts driven locations ample water resources may not be available, in such places water conservation and storage measures are enhanced. The drip irrigation system is another great way to manage water resources without causing any wastage.

Sustainable Forestry

Forests are very important contributors to the environment of the earth. They are not just a habitat for wild life but help in cleaning the air from harmful pollutants prevent soil erosion and contribute to the cloud seed formation. Forests are the main platform for the existence of an ecological system that not only houses it but actually sustains it.

But due to aggressive exploitation by humans have led to deforestation worldwide which has affected the environment by great measures. Sustainable forestry offers a chance to retain some of this natural resource in its original form by ensuring that the numbers of trees that are felled are replaced immediately with healthy saplings to replenish the number of trees in the forest.

In a sustainable forest, seedlings are grown in green houses which are then transferred to the forest. The seedlings then grow up to be young trees in about 10 years within the natural environment. Every 5 years the trees are thinned out to ensure the continued growth of the stronger trees. After 25-30 years the trees are felled and sent to the wood or paper mill. New saplings are then planted to replace the felled trees and the process continues. This is one of the great ways of sustainability and to make our environment more greenery.

Sustainable Construction

The highest amount of energy is consumed every year for domestic purposes worldwide which include heating, cooling and lighting. This can be lowered by a substantial amount by ensuring that the homes, offices and other structures that are built are energy efficient and they incorporate renewable and recycled resources. In a sustainable construction an environmentally responsible and resource efficient process is applied throughout the life cycle of the building starting with the planning, design, the construction method, operation and maintenance.

In colder climates buildings are well insulated and constructed to ensure maximum surface exposure to sunlight and allow sunlight to enter the homes and allow natural warming and lighting of the rooms. In warmer or hotter climates, the buildings are well ventilated and the exposure to sunlight is minimized to allow lower heating but proper lighting. This not only lowers the cost of temperature control but also saves up a lot on the lighting cost.

Incorporation of green spaces into the residential or office areas not only enhances the aesthetics of the location but also provide a great outdoor location to relax and rejuvenate. These small patches of green across a city can go a long way in bringing in a positive effect on our lives and the environment.

Waste Management

Tons of waste is produced everyday in every country. The thoughtless disposal of these wastes into the water bodies since years has had grave results on the environment and the health of people. The aquatic life has had to suffer due to the disposal of waste into the water system. Even if the waste is not disposed in the water but discarded in the dump yards or incinerated then too it poses a great threat to the environment because of the foul smell and the rot it will bring in.

Sustainable waste management starts from the very start of the life cycle of a product. The design and manufacture process in sustainable waste management is developed in a way that it yields the minimum waste and even if it did generate waste then the disposal of the waste like the packaging should not impact the environment or should have minimal effect. Many companies have now started packing the products in biodegradable material which has a minimal impact on the environment and is grey water safe.

Composting is a very effective way to deal with organic waste that can be converted to manure for gardening or farming. The responsible segregation of waste to organic and recyclable forms further helps in ensuring that the wastes like metal, plastic or nylon can be recycled instead of discarding it for incineration or using it for landfill.

One of the greatest threats to the environment is the generation of a large amount of plastic waste every day. It poses a threat not just to the aquatic life but also due to the fact that most of them are not biodegradable the fertility of the soil is impacted. Burn ing of plastic waste is not advisable as well due to the emission of harmful Dioxins that are emitted. The best way to manage the plastic waste is to recycle them. Plastic waste is collected and recycled to make new plastic products or incorporated into other products. In recent years plastic waste has been used in the construction of houses and roads that have been reported to be much stronger than the usual roads and houses made of conventional materials.

Sustainable Water Management

Our planet is covered mostly by water, but what is needed is drinkable and usable water which is a much rarer form. The constant use and disposal of water is leading to a fast depletion of this basic necessity of life which raises a vital question, how long can we continue to use this natural resource to meet our needs?

Various methods of water purification have been developed to make water drinkable and usable. However, it is important we try to use sustainable water management methods to ensure the water supply is not exhausted. The first step towards this is to install water efficient shower heads and low flow taps in our houses to make sure that there is no wastage of water and only the amount that is needed is used. Building complexes, be it commercial or residential can profit largely from a water treatment plant which can help to treat used water from the various users in the building and make it reusable again.

Using water efficient methods of gardening and farming can help in reducing the wastage of water. Using sprinklers and drip water method for farms and garden can help in a great way. Rainwater harvesting has been introduced in many places to ensure that this naturally clean water is not wasted and is stored for future use like drinking or cleaning purposes.

Renewable Energy Resources

The major energy source is the fossil fuels which is fast depleting due to the ever-increasing demand. It is definitely cost effective, but it is not sustainable. Hence it has become very important to find an alternative source of energy which will not only perform in the same way as fossil fuel in generating power but unlike fossil fuel will be sustainable. Here are few of the examples of alternative source of energy to regain a sustainable environment.

- Wind energy: Harnessing the power of wind is a great way to supplement the grid power without causing any harm to the environment. With only an initial investment on setting up a windmill power can be generated free of cost to power offices, buildings, farms or your homes. With the growing awareness of green power technology wind farms have been set up in many places that have good wind speed. The largest wind farm in the world is the Gansu Wind Farm in China with a capacity of over 8GW.

- Solar Energy: The energy from the sun can be harnessed as Solar energy using the various technologies. Solar Energy is the most common form of green power that has become popular all around the world unlike the other renewable resources which are yet to catch up. The photovoltaic system can be set up in an area that is sure to receive the maximum exposure to sunlight throughout the day. The energy can be stored and used as needed. The energy from the sun has been taken advantage of to develop new technologies that would help in replacing the constant need for electricity and other sources of thermal power that leads to environmental hazards. Solar Thermal energy systems have been developed that has been used in effective water heating, cooking, heating homes, and generation of electricity using the molten salt technology. Solar vehicles have been designed to run on solar power. It is not just the development of cars, but fully solar powered boats and airplanes have been developed too. With the advent of newer and advanced technology Solar energy has great promise in replacing the need for fossil fuels.

- Geothermal Energy: The Earth has a very hot core where the temperatures may reach 5000°C. This energy sometimes seeps through the Earth's crust and heats up water on the surface also known as the hot springs or geysers. Geothermal energy has been used for therapeutic baths since ages. It was also used for heating up spaces but more recently it has found use in the generation of electricity.
- Hydropower and Tidal or wave: The hydroelectric power generation is another renewable source of energy which has contributed largely to the generation of electricity. Hydroelectric power has been generated by building dams and reservoirs. More recently Run-of-the-river hydroelectricity has been generated which functions without the creation of reservoirs or dams on rivers and is much more cost effective. China is the largest producer of hydroelectric power in the world. Hydropower is now also being generated by harnessing the power of waves and tides.

Biofuel

This is also a very promising alternative for most fossil fuels. A lot of research is still on going to develop a very efficient form of Biofuel which is not only cost effective but can also be used commercially on vehicles and other machinery just the way fossil fuels are used. Although a few biofuels have been introduced into the market, a lot more work needs to be done towards the impact of cultivation of the biofuel generating crops on the environment. Example of Biofuels that have been tested is organically produced Ethanol from sugarcane, Jatropha seed oil and soya bean oil.

It is very important now that we all get together to save the environment from further depletion. The effects of the thoughtless exploitations by human beings are more apparent now than it was ever before. The climatic changes though started slowly have become more pronounced now and we are experiencing seasons we never did earlier. The catastrophic implications can be avoided by a stringent adherence to environment-friendly methods which will help to sustain the environment for longer.

References

- Menn F, Easter JP, Sayler GS (2001). "Genetically Engineered Microorganisms and Bioremediation". Biotechnology Set. pp. 441–63. doi:10.1002/9783527620999.ch21m. ISBN 978-3-527-62099-9
- Health-effects-toxic-chemicals, whmis-what-you-need-to-know, resources: ehs.utoronto.ca, Retrieved 30 August, 2020
- Brian Shmaefsky (2004). Favorite demonstrations for college science: an NSTA Press journals collection. NSTA Press. p. 57. ISBN 978-0-87355-242-4
- Sigel, A. (2010). Sigel, H.; Sigel, R.K.O. (eds.). Organometallics in Environment and Toxicology. Metal Ions in Life Sciences. 7. Cambridge: RSC publishing. ISBN 978-1-84755-177-1
- Environmental-sustainability-examples, sustainable-development: planningtank.com, Retrieved 27 July, 2020

Applications 7

- Applications of Green Chemistry Principles in Agriculture
- Green Chemistry and Synthesis of Anticancer Molecule

Green chemistry has a wide range of applications in agriculture, medical science and industries. Its major applications include the preservation and maintenance of environment sustainability, using agrochemicals, etc. This chapter sheds light on the various applications of green chemistry for a thorough understanding of the subject.

Applications of Green Chemistry Principles in Agriculture

Green chemistry play a new paradigm in the field of agriculture. Sustainable agriculture and green chemistry are both revolutionary fields and intertwined. In last few years, for sustainable production in agriculture use of renewable biomass resources increases to generate bio-based food products with low inputs, zero waste, substantial social values and minimizing environmental impact.

The problem of poverty in developing countries increases the demand of more productive and industrialized economies, which cause the global and local environmental pollution, and the non-sustainable use of natural resources. Environmental pollution threats, ranging from atmospheric pollution in cities, acid rain, municipal solid waste, deforestation and desertification, the reduction of ozone layer and signs of climate change were overlooked. The idea of sustainable eco-development was presented for the first time in 1987 in the report of the World Commission on Environment and Development of the United Nations.

In last few years, the production of synthetic pescticides was increasing and the modern agriculture methods produced major green house gases e.g. emission of 84 % nitrous oxide (N_2O) annually in all over the world.

The adverse impact of contamination in agrochemical fields through indirect or direct exposure of improper use of pesticides, effects the animals and human health. Pesticides includes all chemicals which are used to control or kill the pests, but these pesticides in food chain coupled with biomagnifications and bioaccumulation having adverse effects on entire animals and human life. To minimize these harmful effects, employment of organic farming should be increased in place of synthetic pesticides.

These pesticides also causes the contamination of ground water and led to eutrophication of rivers and lake waters, and the movement of toxic chemicals from surroundings into an organism. There are some pesticides e.g. DDT, which are soluble and accumulate in fatty tissues and causes biomagnifications in food chain.

Green chemistry involves the design and development of products and processes that minimize or eliminate the use and generation of chemicals hazardous to the environment and human health. The principles of green chemistry involve the development of green catalysts and use of non-toxic reagents. Green chemistry emphasizes the use of reactions improved atom efficiency, use of solvent free or environmentally benign recyclable solvent system and use of renewable resources.

In 1998, Paul Anastas and John C. Warner developed the 12 principles of green chemistry. Green chemistry starts with molecular level and eventually indicating the most important environmentally benign processes and products. In 2001, at University of Colorado, Boulder, IUPAC CHEMRAWN XIV Conference on Green Chemistry: Towards Environmentally Benign Processes and Products, was organized and in this conference a number of chemists discuss the effects of agricultural and industrial activities on atmospheric chemistry. They advised "design-for-environment" framework to work with consumers and assure the safety of foods and crops developed by green methods for green agricultural practice.

Nowadays, green chemistry play a new paradigm in the field of agriculture. In last few years, for sustainable production in agriculture use of renewable biomass resources increases to generate bio-based food products with low inputs, zero waste, substantial social values and minimizing environmental impact. Sustainability of agriculture is the core area which requires green chemistry strategies in agrochemical field for implanting the judicious use of pesticide and fertilizers.

Green Principles Application in Agriculture

- Beginning with the selection of renewable, non-toxic feedstocks.
- The design of safe and energy efficient synthetic procedures.
- Maximum incorporation of all materials into the product, eliminating auxiliaries when possible.
- Generating durable, non-toxic products with preserved function.

- Ensuring the natural degradability of all products and by-products at the end of life.
- The principles of green chemistry are especially relevant to the manufacturing of agrichemicals due to their direct impact on human and environment health. However, current agricultural practices are still based on intensive production methods using unsustainable technologies developed during the 'green revolution.
- This technology is characterized by the extensive use of high yielding crop varieties, chemical fertilizers, pesticides and irrigation. According to market research by FoodThink, 66% of Americans feel the agriculture industry is not transparent about food production practices citing primary concerns of the use of pesticides and insecticides, animal antibiotics, and animal hormones.
- As consumer focus shifts towards establishing a sustainable and secure food supply, the agrichemical industry will require a second 'green revolution' utilizing green chemistry principles to continue providing products relevant to agricultural practices.
- The bio pesticides itself has scalable and green technology which has direct implication on sustainable agriculture for broad range production. The innovative work used for the preparation of biocatalysis processes using reductases, transaminaes, oxidases, hydrolases etc. Nowadays, innovative enzymes are used for synthesis of biocatalysts which offered economical benefits in pharmaceutical and agriculture.
- Nowaday, the manufacturer emphasizing the replacement of conventional solvents to green solvents for worker safety in chemical and environmental protection. The selection of right solvents has always increase the benefits to environment, safer to user and improved awareness. Currently, the green solvents represented the around 10% market. In the last few years, laboratory synthesis and manufacturing in industry replacing into catalyst free reactions and catalyst free reactions on water and in water.
- Recently, organic solvents with high volatility and inherent toxicity were change by the ionic liquids (ILs). Ionic liquids are organic salts usually melt below 1000C, high thermal stability, nearly non-volatility under normal conditions, dissolve non-polar and polar organic, inorganic compounds. For these reason, ionic liquids called "designer solvents".
- The renewable energy resources such as solar, wind, hydroelectric, biomass, biorefifineries, geothermal and ocean energy are important resources for future sustainable development, so it will replace carbon containing sources and reduces the emission of global warming. In the last years, third-generation biofuels (high yielding-low input feedstock) derived from renewable feedstock.

- There are 234 registered pesticides. Out of them, 4 are WHO class Ia pesticides, 15 are WHO class Ib pesticides and 76 are who class II pesticides.
- Recently, a large number of deaths of farmers increases due to pesticides poisoning. There is a list of highly or extremely hazardous seven pesticides which are continued to be used despite of being banned in many countries. Pesticides like Monocrotophos and Oxydemeton-methyl are considered Class-I pesticides by WHO. "Since Class-I pesticides can be fatal at a very low dose, many of these are banned in several countries", but allowed to sold and still use of these pesticides in some area of countries, which cause death of farmers. According to official estimate, death of at least 10,000 people every year due to direct pesticide poisoning.
- As we see some of the chemicals still used conventionally in agriculture which are associated with adverse impacts to environment and human health. There are growing concern towards about how we farm for sustainable agriculture, what inputs would supply to it, what technologies to employ. The green chemistry will shift the agriculture towards sustainable development.

Green chemistry seeking the goal towards farm profitability, community prosperity and improving soil quality by reducing the dependence use of non-renewable resources e.g synthetic fertilizers and pesticides, minimizing the adverse effects on water quality, wildlife and safety. There are various alternatives of chemical farming such as biological agriculture, organic farming, natural farming, bio-dynamic agriculture, ecological agriculture. Bio-pesticides are organic in nature, so these can be employed in farming for controlling the pests, insects, weeds and also for plant physiology and productivity. These bio pesticides are bio-degradable to the environment. Therefore, for sustainable developments, shift the agricultural farming into green chemistry manufacturing processes, use of crop protection and production and develop green agrochemicals.

Thus, sustainable agriculture and green chemistry are both revolutionary fields and intertwined. For this green chemists needs that the farmers used green technology for sustainable agriculture and farmers needs safe and green agricultural inputs. The biocatalysts have been used increasingly in agrochemicals, pharmaceuticals and food industries, these can help in reducing the waste and improving the yield of products.

The population of world stands at 7 billion and increases upto 9.3 billion by 2050. This will required food production to be increased by 70% in order to meet the demand. Therefore, it is necessary to increases the crop productivity, reduced preharvest loss and post harvest lost from pest attack by employing biochemical processes with green techniques.

The agrochemical industries have carrying major growth in futures. The industry, regulatory bodies, government and academia can build a collaborative work together and employs their ideas of green chemistry practices.

Agrochemicals

An agrochemical is any substance that humans use to help in the management of an agricultural ecosystem. Agrochemicals include fertilizers, liming and acidifying agents (which are designed to change the pH), soil conditioners, pesticides, and chemicals used in the raising of livestock such as antibiotics and hormones.

The use of agrochemicals is an increasingly prominent aspect of modern agriculture. As farms have become massive in size, the challenges in keeping the crop free of damage have increased. Hand-tilling weeds has become impractical, as one example. Thus, agrochemicals have become widely used.

The use of agrochemicals has been critically important in increasing the yield of agricultural crops. However, some uses of agrochemicals cause environmental and ecological damage, which detracts significantly from the benefits gained by the use of these materials.

Fertilizers

Fertilizers are substances that are added to agricultural lands to alleviate nutrient deficiencies, allowing large increases in the rates of crop growth. As of 2006, over 150 million tons of fertilizer are used in agriculture each year worldwide. In the United States, the average rate of fertilizer application is about 220 pounds per 2.5 acres (approximately 100 kilograms per hectare), and about 21 million tons (19 million metric tons) are used in total each year.

The most commonly used fertilizers are inorganic compounds of nitrogen (N). Under conditions where agricultural plants have access to sufficient water, their productivity is most often constrained by the supply of available forms of nitrogen, especially nitrate (NO_3^-), and sometimes ammonium (NH_4^+). Farmers commonly increase the availability of these inorganic forms of nitrogen by applying suitable fertilizers, such as urea or ammonium nitrate. The rate of fertilization in intensive agricultural systems is commonly several hundred pounds of nitrogen per acre per year.

Phosphorus (P) and potassium (K) are other commonly applied nutrients in agriculture. Most phosphorus fertilizers are manufactured from rock phosphate, and are known as superphosphate and triple-superphosphate. Some other phosphorus fertilizers are made from bone meal or seabird guano. Potassium fertilizers are mostly manufactured from mined potash.

Often, these three macronutrients are applied in a combined formulation that includes nitrogen, in what is known as an N-P-K fertilizer. For example, a 10-10-10 fertilizer would contain materials equivalent to 10% each of nitrogen (N), P_2O_5 (source of phosphorus [P]), and K_2O (source of potassium [K]), while a 4-8-16 fertilizer would contain these nutrients in concentrations of 4%, 8%, and 16%, respectively. The desired ratios of these three nutrients are governed by the qualities of the soil being fertilized, and by the needs of the specific crop.

Sometimes other nutrients must also be supplied to agricultural crops. Sulfur, calcium, or magnesium, for example, are limiting to crop productivity in some places. Rarely, micronutrients such as copper, molybdenum, or zinc must be applied to achieve optimum crop growth.

Liming and Acidifying Agents

Agricultural soils are commonly too acidic or too alkaline for the optimal growth of many crop species. When this is the case, chemicals may be added to the soil to adjust its pH to a more appropriate range. Acidic soils are an especially common problem in agriculture. Acidic soil can be caused by various factors, including the removal of acid-neutralizing bases contained in the biomass of harvested crops, the use of certain types of fertilizers, acid rain, the oxidation of sulfide minerals, and the presence of certain types of organic matter in soil. Because soil acidification is such a common occurrence, acid-neutralizing (or liming) materials are among the most important agrochemicals used, in terms of the quantities added to soil each year.

Acidic soils are commonly neutralized by adding calcium-containing minerals, usually calcite ($CaCO_3$) in the form of powdered limestone or crushed oyster or mussel shells. Alternatively, soil acidity may be neutralized using faster-acting lime ($Ca[OH]_2$). The rate of application of acid-neutralizing substances in agriculture can vary greatly, from several hundred pounds per acre per year to more than 1,000 pounds per acre per year. The rates used depend on the acidity of the soil, the rate at which new acidity is generated, and the needs of specific crops.

Much less commonly, soils may be alkaline in reaction, and they may have to be acidified somewhat to bring them into a pH range suitable for the growth of most crops. This problem can be especially common in soils developed from parent materials having large amounts of limestone ($CaCO_3$) or dolomite ($CaMg[CO_3]_2$). Soil can be acidified by adding sulfur compounds, which generate acidity as they are oxidized, or by adding certain types of acidic organic matter, such as peat mined from bogs.

Soil Conditioners

Soil conditioners are organic-rich materials that are sometimes added to soils to improve aeration and water-holding capacity, both of which are very important aspects of soil quality. Various materials can be utilized as soil conditioners, including peat, crop residues, livestock manure, sewage sludge, and even shredded newspapers. However, compost is the most desirable of the soil conditioners. Compost contains large quantities of well-humified organic compounds, and also supplies the soil with nutrients in the form of slow-release organic compounds.

Pesticides

Pesticides are agrochemicals that are used to reduce the abundance of pests, that is, organisms that are considered to interfere with some human purpose. Many kinds of pesticides are used in agriculture, but they can be categorized into simple groups on the basis of the sorts of pests that are the targets of the use of these chemicals. Herbicides are used to kill weeds, that is, non-desired plants that interfere with the growth of crops and thereby reduce their yield. Fungicides are used to protect agricultural plants from fungal pathogens, which can sometimes cause complete failure of crops. Insecticides are used to kill insects that defoliate crops, or that feed on stored grains or other agricultural products. Acaricides (or miticides) are used to kill mites, which are pests of crops such as apples, and ticks, which can carry debilitating diseases of livestock. Nematicides are used to kill nematodes, which are parasites of the roots of some crop species. Rodenticides are used to kill rats, mice, gophers, and other rodents that are pests in fields or that eat stored crops. Preservatives are agrochemicals added to processed foods to help prevent spoilage.

Pesticides are chemically diverse substances. About 300 different insecticides are now in use, along with about 290 herbicides, 165 fungicides, and other pesticides. However, each specific pesticidal chemical (also known as the "active ingredient") may be marketed in a variety of formulations, which contain additional substances that act to increase the efficacy of the actual pesticide. These so-called "inert" ingredients of the formulation can include solvents, detergents, emulsifiers, and chemicals that allow the active ingredient to adhere better to foliage. In total, more than 3,000 different pesticide formulations exist.

Pesticides can also be classified according to the similarities of their chemical structures. Inorganic pesticides, for example, are simple compounds of toxic elements such as arsenic, copper, lead, and mercury. Inorganic pesticides were formerly used in large quantities, especially as fungicides. However, they have largely been replaced by various organic (carbon-containing) pesticides. A few of the commonly used organic pesticides are based on substances that are synthesized naturally by plants as biochemical defenses, and can be extracted and used against pests. Pyrethrin, for example, is an insecticide based on pyrethrum, which is obtained from a species of chrysanthemum, while rotenone is a rodenticide extracted from a tropical shrub.

Most organic pesticides, however, have been synthesized by chemists. The synthetic organic pesticides include such well-known groups as the chlorinated hydrocarbons (including the insecticide DDT, and the herbicides 2, 4-D and 2, 4, 5-T), organophosphates (such as parathion and malathion), carbamates (for example, carbaryl and carbofuran), and triazine herbicides (such as atrazine and simazine).

A final class of pesticides is based on the action of bacteria, fungi, or viruses that are pathogenic to specific pests and can be applied as a pesticidal formulation. The most commonly used biological insecticide is manufactured using spores of the bacterium Bacillus thuringiensis, also known as Bt. These spores can be mass-produced in laboratory-like factories, and then used to prepare an insecticidal solution. Insecticides containing Bt are mostly used against leaf-eating moths, biting flies, such as blackflies, and mosquitoes. Most other insects are little affected by Bt-based insecticides, so the unintended nontarget effects of their usage are relatively small.

Agrochemicals used for Animal Husbandry

Contagious diseases of livestock can be a problem in modern agriculture. This is especially true when animals are being reared at a high density, for example, in feedlots. Various agrochemicals may be used to control infectious diseases and parasites under such conditions. Antibiotics are especially important in this respect. These chemicals may be administered by injection whenever bacterial diseases are diagnosed. However, antibiotics are sometimes administered with the feed, as a prophylactic treatment to prevent the occurrence of infections. Because of the extremely crowded conditions when livestock are reared in "factory farms," antibiotics must be administered routinely to animals raised under those circumstances. This practice is contentious, as evidence suggests that the development of antibiotic resistant bacteria may be encouraged.

Sometimes, hormones and other animal-growth regulators are used to increase the productivity of livestock. For example, bovine growth hormone is routinely administered in some agricultural systems to increase the growth rates of cows and their milk production. This use is also contentious; critics contend the unclear health effect of the hormone in humans argues against growth hormone supplementation.

Environmental Effects of the use of Agrochemicals

Many important benefits are achieved by the use of agrochemicals. These are largely associated with increased yields of plant and animal crops, and less spoilage during storage. These benefits are substantial. In combination with genetically improved varieties of crop species, agrochemicals have made important contributions to the successes of the "green revolution." This has helped to increase the food supply for the rapidly increasing population of humans on Earth.

However, the use of certain agrochemicals has also been associated with some important environmental and ecological damages. Excessive use of fertilizers, for example, can lead to the contamination of groundwater with nitrate, rendering it unfit for consumption by humans or livestock. Water containing large concentrations of nitrate can poison animals by immobilizing some of the hemoglobin in blood, reducing the ability to transport oxygen. In addition, the run-off of agricultural fertilizer into streams, lakes, and other surface waters can cause an increased productivity of those aquatic ecosystems, a problem known as eutrophication. The ecological effects of eutrophication can include an extensive mortality of fish and other aquatic animals, along with excessive growth of nuisance algae, and an off-taste of drinking water.

The use of pesticides can also result in environmental problems. Pesticides are used in agriculture to reduce the abundance of species of target pests to below a level of acceptable damage, which is economically determined. Unfortunately, during many uses of pesticides in agriculture, the exposure of other organisms, including humans, is not well controlled. This is especially true when entire fields are sprayed, for example, when using application equipment drawn by a tractor, or mounted on an airplane or helicopter. During these sorts of broadcast applications, many non-target organisms are exposed to the pesticide. This occurs on the treated site, and also on nearby off-sites as a result of "drift" of the sprayed agrochemical. These non-target exposures cause many unnecessary poisonings and deaths of organisms that are not agricultural pests.

In addition, there is a widespread, even global contamination of the environment with some types of persistent pesticides, especially with organochlorines such as DDT, dieldrin, and aldrin. This contamination involves the widespread presence of pesticide residues in virtually all wildlife, well water, food, and even in humans. Residues of some of the chemicals used in animal husbandry are also believed by some people to be a problem, for example, when traces of antibiotics and bovine growth hormones occur in consumer products such as meat or milk.

Some of the worst examples of environmental damage caused by pesticides have been associated with the use of relatively persistent chemicals, such as DDT. Most modern usage of pesticides involves chemicals that are less persistent than DDT and related chlorinated hydrocarbons. However, severe damages are still caused by the use of some newer pesticides. In North America, for example, millions of wild birds have been killed each year as a non-target effect of the routine use of carbofuran, an agricultural insecticide. This is a substantial ecological price to pay for the benefits associated with the use of that agrochemical.

The use of some pesticides is also risky for humans. About one million pesticide poisonings occur globally every year, resulting in 20,000 fatalities. About one-half of the human poisonings occur in poorer, less-developed countries, even though these places account for only 20% of the world's use of pesticides. This disproportionate risk is due to greater rates of illiteracy in poorer countries, and to lax enforcement of regulations concerning the use of pesticides.

There have been a few examples of pesticides causing extensive toxicity to humans. The most famous case occurred at Bhopal, India, in 1984, in the vicinity of a factory that was manufacturing an agricultural insecticide. In that case, there was an accidental release of about 45 tons (40 tonnes) of deadly methyl isocyanate vapor to the atmosphere.

This agrochemical-related emission caused the deaths of about 3,000 people, and more than 20,000 others were seriously injured.

These and other environmental effects of the use of some agrochemicals are unfortunate consequences of the application of these chemical tools to deal with agricultural problems. Researchers are constantly searching for non-chemical ways of dealing with many of these agricultural needs. Much attention is being paid, for example, to developing "organic" methods of enhancing soil fertility and dealing with pests. Unfortunately, economically effective alternatives to most uses of agrochemicals have not yet been discovered. Consequently, modern agricultural industries will continue to rely heavily on the use of agrochemicals to manage their problems of fertility, soil quality, and pests.

Green Chemistry and Synthesis of Anticancer Molecule

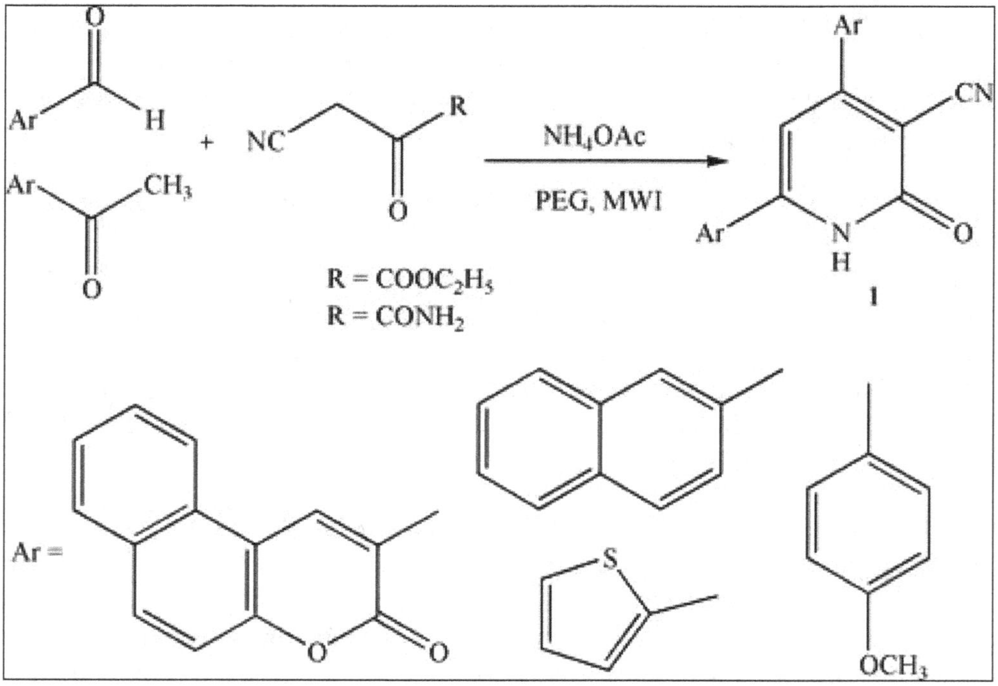

3-cyano pyridine derivatives.

Green chemistry is one of the valuable concepts for the development of new, more effective, solvent-free less toxic, environmentally friendly and cost-efficient methods for the synthesis of different anticancer molecules. There are many developments for the environmentally friendly approaches for the synthesis of biologically active molecules such as microwave-assisted synthesis, one-pot synthesis, solvent-free synthesis, enzyme-catalyzed synthesis, solid phase synthesis, ultrasound promoted and catalyst-free synthesis.

One-pot Synthesis of 3-cyano Pyridine Derivatives

Novel series of 3-cyano pyridine type derivatives were synthesized and their cytotoxic activity was evaluated against many human MCF-7, HCT-116 and HepG-2 cancer cell lines. Most of the compounds showed good-to-moderate activity against HepG2 and HCT-116 cell lines, whereas only few compounds showed significant cytotoxic activity against MCF-7 breast cancer cell line.

Microwave-assisted Solvent-free Synthesis of Stigmastane Derivatives

The microwave-assisted synthesis in most cases was more successful regarding to the reaction time and the yields of product. These reactions are more environmentally friendly too, compared to the conventional synthetic methods. a convenient simple microwave-assisted solvent-free synthesis of 2-methoxybenzoyloxy androstane, salicyloyloxy stigmastane derivatives from methyl salicylate and appropriate steroidal precursors has done. 2-Methoxybenzoyl ester exhibited significant cytotoxic activity against MDA-MB-231 cells. Most of the compounds strongly inhibited growth of PC-3 cells, whereas salicyloyloxy stigmastane derivative showed the best inhibition potency.

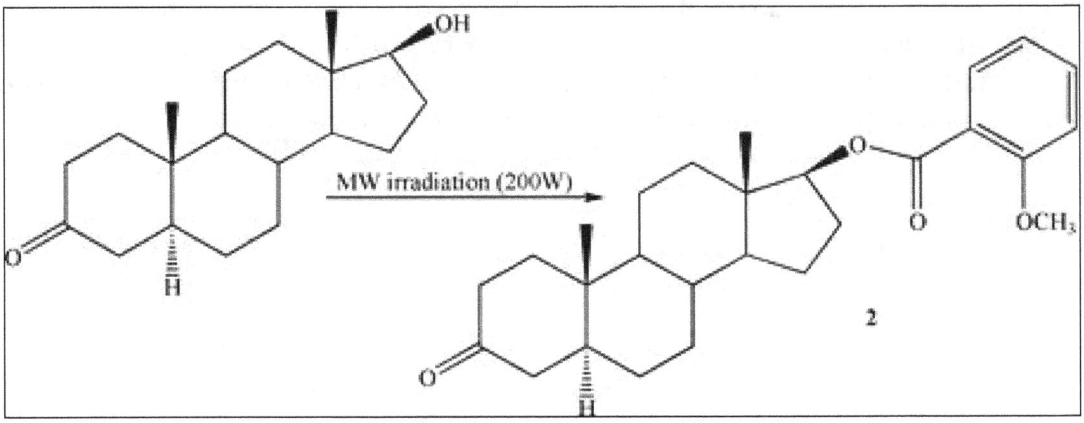

Synthesis of stigmastane derivatives.

One-pot Synthesis of Polyazaheterocycles in Water

Synthesis of these polyazaheterocycles was carried out by green synthetic strategy that involved one-pot azidation and CuAAC under mild conditions in water. Many compounds were synthesized and evaluated for their cytotoxic effects against four human cancer cell lines, including A549 (lung), MCF-7 (breast), HepG2 (hepatocellular) and DU145 (prostate). Some of the compounds showed strong activities against A549 cancer cells.

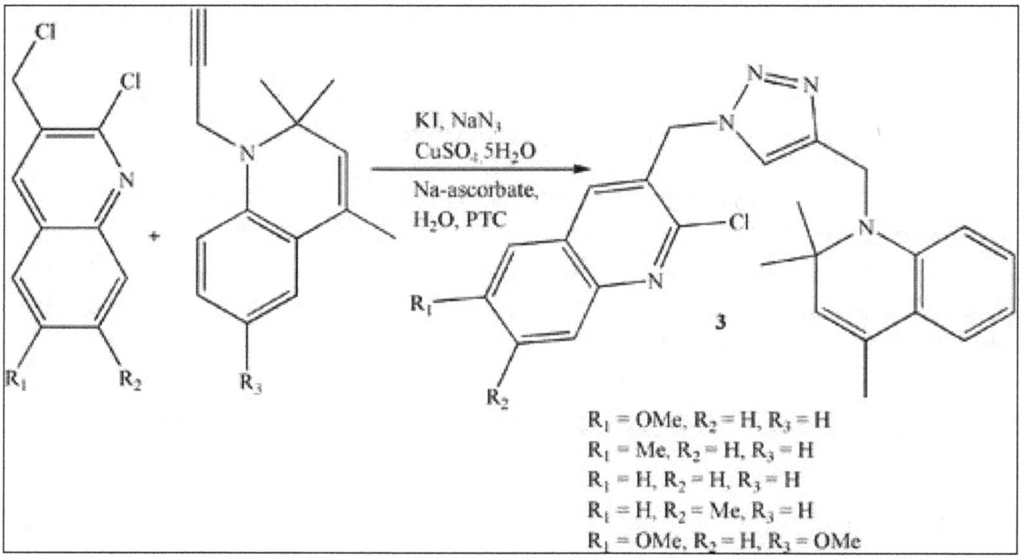

Synthesis of polyazaheterocycles.

Microwave Irradiated One-pot Synthesis of Carboethoxy/Carbome-thoxy Derivatives

Fourteen carboethoxy/carbomethoxy derivatives have been synthesized by conventional and microwave irradiation method from a one-pot three-component reaction mixture, consisting of, alkyl acetoacetate, aldehyde and ammonium acetate. The synthesized products have been evaluated for their cytotoxic activity against MDA-MB (breast) and HT-29 (colon) human cancer cell lines. Few compounds exhibit some degree of cytotoxicity and it was low when compared with standard.

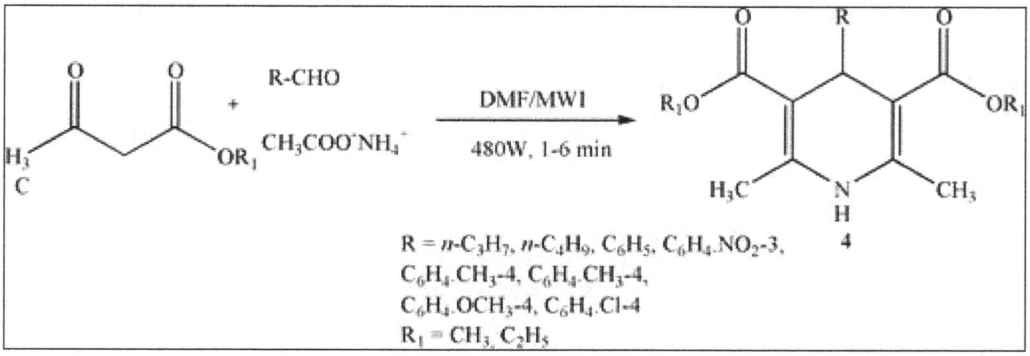

Synthesis of carboethoxy/carbomethoxy derivatives.

Cellulose-supported Copper Nanoparticle-catalyzed Synthesis of Chalcone Derivatives

Chalcone-linked 1,2,3-triazole derivatives were synthesized in water by cellulose-supported copper nanoparticle-catalyzed click reaction. All the products were subjected to MTT cytotoxicity assay against four human cancer cell lines A549, MCF-7, HepG2 and

MIA-Pa-Ca-2 for testing their anticancer potential. Few compounds were found to be most active against all cancer cell lines and showed better activity when compared to reference drug.

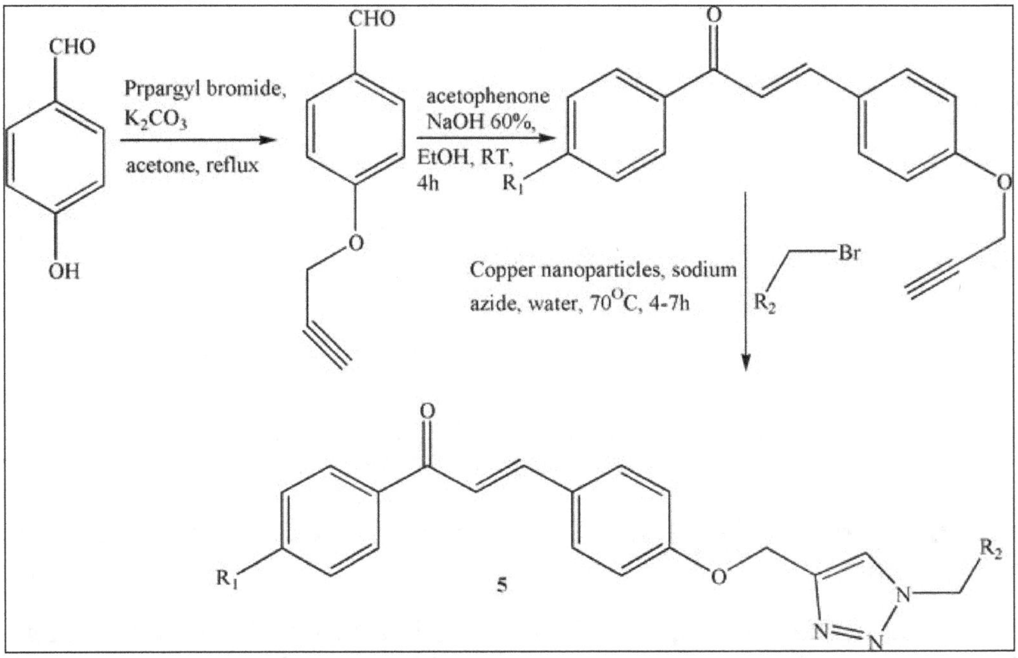

Synthesis of chalcone derivatives.

Copper-mediated Synthesis of Pyrazolo Pyridine Derivatives

Some novel pyrazolo pyridine type compounds were synthesized by facile procedures and showed significant cytotoxic potential on different human cancer cell lines. They revealed various cancer cell lines (HCT-116, A549, MCF-7, PC-3) determined by SRB assay.

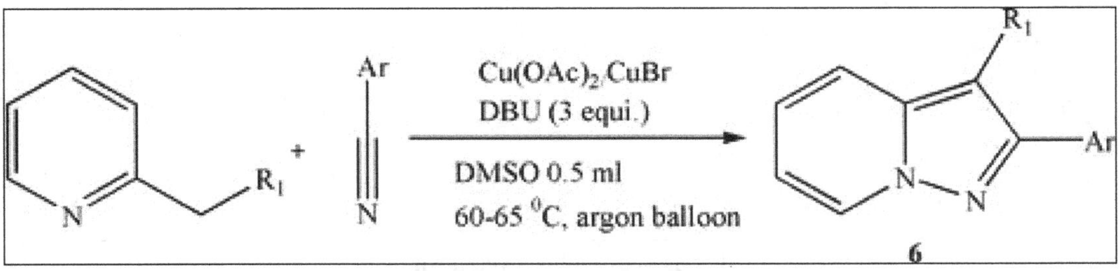

Synthesis of pyridine derivatives.

Microwave-assisted Synthesis Of Quinoline Analogues

A new class of pyrazolo[4,3-c]quinoline and pyrano[3,2-c]quinoline analogues was synthesized in good yields by microwave conditions. For enhancing the yield of products, multicomponent one-pot synthesis has been developed. The cytotoxicity of these compounds was also evaluated against MCF-7 and A549 cancer cell lines. Most of the compounds displayed moderate-to-good anticancer activity against these cell lines.

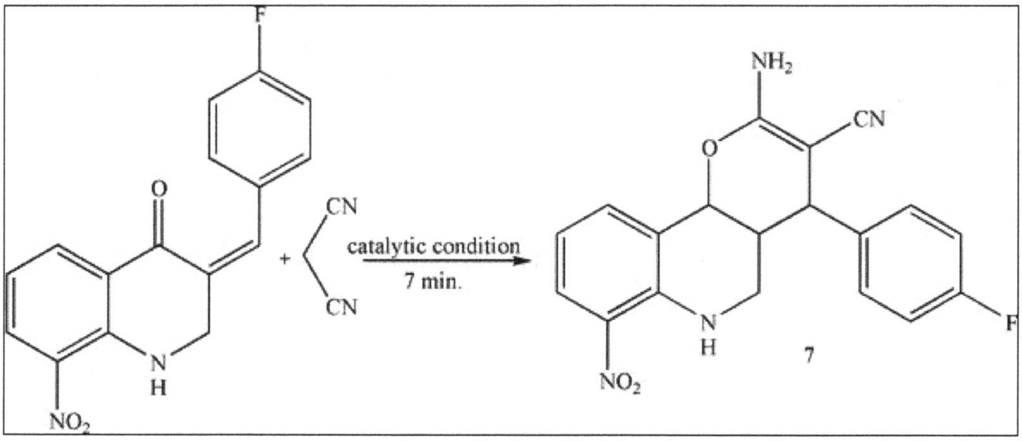

Synthesis of quinoline analogues.

Facile Protocol, Efficient and Environmentally Benign Synthesis of Cycloheximide

In this research, they describe a facile and efficient protocol and environmentally benign for the synthesis of C_5-substituted barbiturate acid in water. The synthesized compounds tested for different assay and provided promising results against a-glucosidase inhibitor. The cytotoxic activity of compound against 3T3 cell resulted that compounds showed significant to weak activity against the standard cycloheximide.

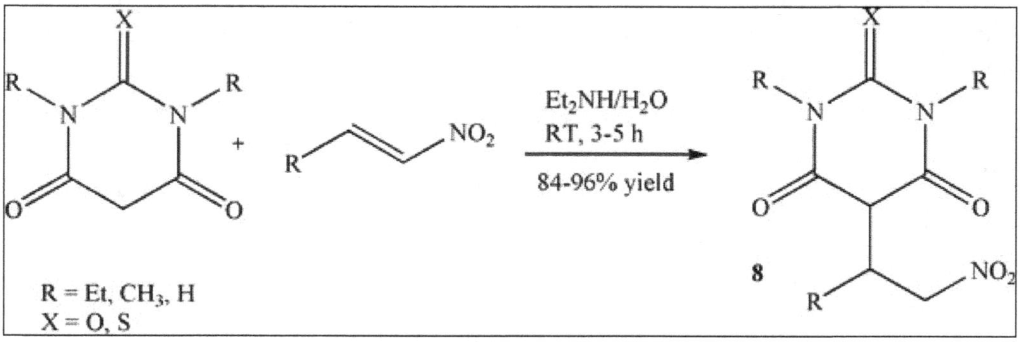

Synthesis of quinoline cycloheximide.

Simple, Eco-friendly and Efficient Synthesis of α,β-unsaturated Carbonyl Compounds

A novel series of carbonyl compounds was synthesized by environment-friendly, simple and efficient method. Compounds were tested for cytotoxicity. All strong antioxidant compounds showed strong protective effect against PC12 cell line.

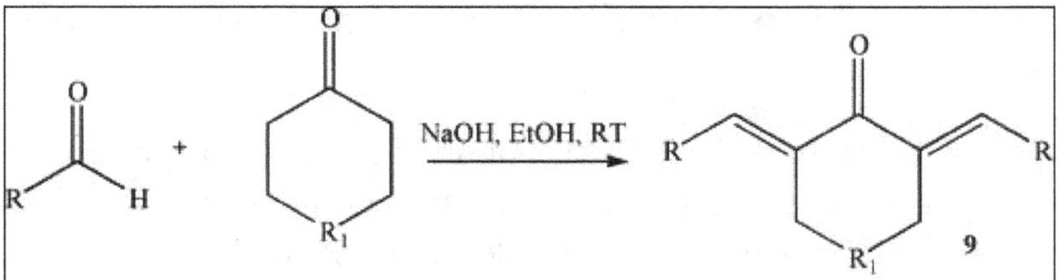

R	R_1
4-Diethoxymethyl-benzene	CH_2
4-Diethoxymethyl-benzene	O
4-Diethoxymethyl-benzene	NH
4-Diethoxymethyl-benzene	NCH_3
4-Diethoxymethyl-benzene	NCH_2-benzene
Phenyl-2-NO_2-4-diethylamine	CH_2
Phenyl-2-NO_2-4-diethylamine	O
Phenyl-2-NO_2-4-diethylamine	NH
Phenyl-2-NO_2-4-diethylamine	NCH_3
Phenyl-2-NO_2-4-diethylamine	NCH_2-benzene
Phenyl-4-pyrrolidine	CH_2
Phenyl-4-pyrrolidine	O
Phenyl-4-pyrrolidine	NH
Phenyl-4-pyrrolidine	NCH_3
Phenyl-4-pyrrolidine	NCH_2-benzene

Synthesis of α,β-unsaturated carbonyl-based compounds.

Green Methodology Synthesis of 2H-indazolo[2,1-b]phthalazine-trione Derivatives

An efficient green method was used for the synthesis of 2H-indazolo[2,1-b]phthalazine-trione derivatives. Many compounds were obtained in good yields within 10 min. Among all tested cell lines, K562 leukemia cell line was most sensitive.

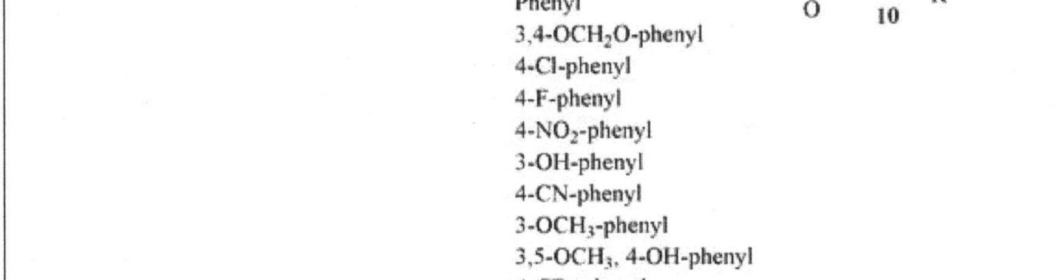

R
Phenyl
3,4-OCH_2O-phenyl
4-Cl-phenyl
4-F-phenyl
4-NO_2-phenyl
3-OH-phenyl
4-CN-phenyl
3-OCH_3-phenyl
3,5-OCH_3, 4-OH-phenyl
4-CF_3-phenyl
4-SCH_3-phenyl
n-propyl
n-butyl
H

Synthesis of phthalazine-trione derivatives.

Green Synthesis of Thioxoimidazolidin and Benzothiazole Derivatives

A series of 2-thioxoimidazolidin-4-one and benzothiazole thioglycosides were synthesized by one-pot reaction. The cytotoxic activity of compound was evaluated against MCF-7 breast cell and it showed high-to-moderate anticancer activities.

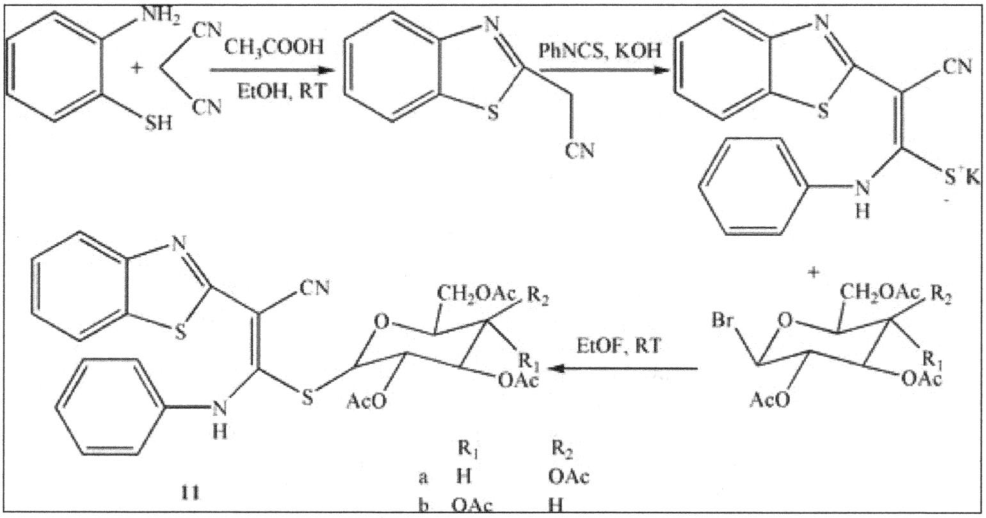

Synthesis of thioxoimidazolidin and benzothiazole derivatives.

Green Synthesis of Pyrimidine-6-carbonitrile Derivatives

This is a green synthetic approach for the formation of antitumor active 5-amino-2-(4-chlorophenyl)-7-substituted phenyl-8,8a-dihydro-7H-(1,3,4)thiadiazolo (3,2-α) pyrimidine-6-carbonitrile. This protocol is extendable to a wide variety of many substrates. The advantages are the use of eco-friendly catalyst, reduced time, simple work-up process, ease of isolation and high yield of product. One compound was found to have the highest GI50 value for PC-3, HeLa, K562 and MCF-7 cancer cell lines.

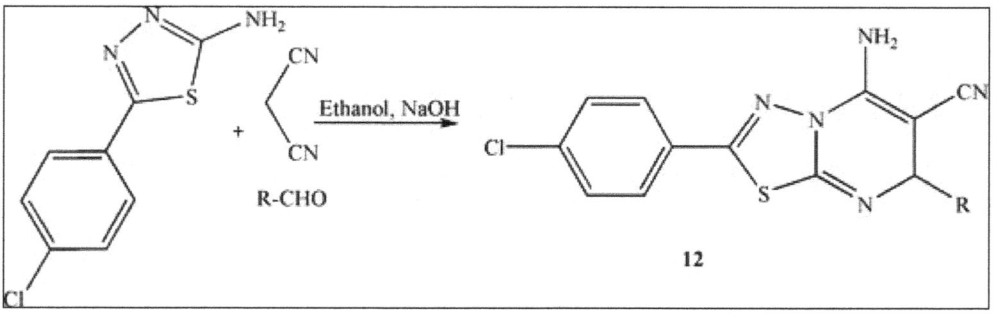

Synthesis of pyrimidine-6-carbonitrile derivatives.

Microwave-assisted Synthesis of Benzimidazole Derivatives

Twelve 2-quinolizinylbenzimidazole and 2-naphthalylbenzimidazole type compounds have been synthesized under MW microwave condition. These compounds were tested for cytotoxicity against human breast cancer cell line MCF-7. The results showed that some compounds were found to be as active as standard Tamoxifen.

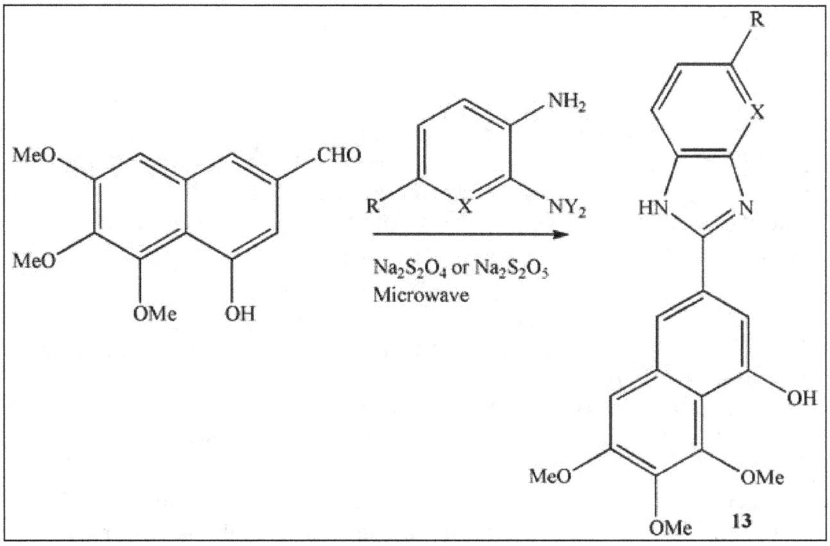

Synthesis of pyrimidine-6-carbonitrile derivatives.

PERMISSIONS

All chapters in this book are published with permission under the Creative Commons Attribution Share Alike License or equivalent. Every chapter published in this book has been scrutinized by our experts. Their significance has been extensively debated. The topics covered herein carry significant information for a comprehensive understanding. They may even be implemented as practical applications or may be referred to as a beginning point for further studies.

We would like to thank the editorial team for lending their expertise to make the book truly unique. They have played a crucial role in the development of this book. Without their invaluable contributions this book wouldn't have been possible. They have made vital efforts to compile up to date information on the varied aspects of this subject to make this book a valuable addition to the collection of many professionals and students.

This book was conceptualized with the vision of imparting up-to-date and integrated information in this field. To ensure the same, a matchless editorial board was set up. Every individual on the board went through rigorous rounds of assessment to prove their worth. After which they invested a large part of their time researching and compiling the most relevant data for our readers.

The editorial board has been involved in producing this book since its inception. They have spent rigorous hours researching and exploring the diverse topics which have resulted in the successful publishing of this book. They have passed on their knowledge of decades through this book. To expedite this challenging task, the publisher supported the team at every step. A small team of assistant editors was also appointed to further simplify the editing procedure and attain best results for the readers.

Apart from the editorial board, the designing team has also invested a significant amount of their time in understanding the subject and creating the most relevant covers. They scrutinized every image to scout for the most suitable representation of the subject and create an appropriate cover for the book.

The publishing team has been an ardent support to the editorial, designing and production team. Their endless efforts to recruit the best for this project, has resulted in the accomplishment of this book. They are a veteran in the field of academics and their pool of knowledge is as vast as their experience in printing. Their expertise and guidance has proved useful at every step. Their uncompromising quality standards have made this book an exceptional effort. Their encouragement from time to time has been an inspiration for everyone.

The publisher and the editorial board hope that this book will prove to be a valuable piece of knowledge for students, practitioners and scholars across the globe.

INDEX

www.ingramcontent.com/pod-product-compliance
Lightning Source LLC
LaVergne TN
LVHW080627210926
848485LV00003B/226